Biodiversity Offsets

Wolfgang Wende • Graham - M. Tucker •
Fabien Quétier • Matt Rayment • Marianne Darbi
Editors

Biodiversity Offsets

European Perspectives on No Net Loss of Biodiversity and Ecosystem Services

Springer

Editors
Wolfgang Wende
Leibniz Institute of Ecological Urban and Regional Development and TU Dresden
Dresden, Germany

Graham - M. Tucker
Institute for European Environmental Policy
London, UK

Fabien Quétier
BIOTOPE
Mèze, France

Matt Rayment
Consultant
London, UK

Marianne Darbi
Helmholtz Centre for Environmental Research
Leipzig, Germany

ISBN 978-3-030-10230-2 ISBN 978-3-319-72581-9 (eBook)
https://doi.org/10.1007/978-3-319-72581-9

Printed on acid-free paper

This Springer imprint is published by Springer Nature
The registered company is Springer International Publishing AG
The registered company address is: Gewerbestrasse 11, 6330 Cham, Switzerland

Preface

The intensification of land use and increasing urbanisation seen in Europe are causing a loss of biodiversity and ecosystem services on a dramatic scale. In response, Europe has successfully established Natura 2000, a Europe-wide network of protected sites, and protected many species of threatened fauna and flora. This is certainly an important step in the right direction. Yet outside of protected areas and beyond protected species, there are so far—at least on a pan-European level—few measures to reduce the loss of biodiversity and associated ecosystem services, and this despite the fact that the intensification of land use shows no sign of letting up. This policy gap in the wider environment is also a reason why the decline and loss of species has not (yet) been stopped in the EU.

Outside Europe, in countries as diverse as the USA, Australia, Brazil and South Africa, considerable efforts are being made to introduce and implement policies in the wider environment that strengthen or expand requirements to avoid, mitigate and, if necessary, offset impacts on biodiversity and ecosystem services. Offsets implement the "polluter pays" principle and thereby incentivise the reduction of impacts and require no net loss (or a net gain) to be achieved where residual impacts are unavoidable. With its Wetland Mitigation programme, the USA has long been a pioneer in this field. Australia can point to its Bushbroker scheme set up several years ago, while compensation instruments have also been introduced in South Africa. Brazil has its own system to address deforestation in private land and access private funding for biodiversity conservation and ecosystem restoration. There is now a lively international debate among practitioners and scientists alike on the optimal design of mitigation hierarchies and offset systems to ensure their success and to minimise the risks of the failure or misuse of offsets.

This debate is also taking place in Europe. However, offsetting in Europe needs to take into account the continent's unique natural heritage, where some of its current biodiversity is associated with habitats that have been modified by human activities and culture. As a contribution to this debate, this book aims to present the activities and practices of biodiversity offsetting already under way in selected EU member states and the lessons that can be learnt from them. Readers may be surprised at how much experience already exists in these countries. It is also notable that in many countries, the focus of offsetting policies is justified by the intrinsic value of biodiversity, including threatened species and habitats. But some countries have also considered a more diverse set of values tied to nature, including people's sense of place, or the historical or cultural values tied to certain features. Building on this experience, and recent scientific developments, there is a trend towards broadening the scope of offsetting in Europe by considering natural capital and ecosystem services. A further aim of the book is to offer grounded insights on the road ahead and foster a more intensive and fruitful discussion on how offsetting can be extended and improved upon, so that it becomes a key and effective component of Europe's biodiversity conservation policy framework.

Dresden, Germany	Wolfgang Wende
London, UK	Graham - M. Tucker
Mèze, France	Fabien Quétier
London, UK	Matt Rayment
Leipzig, Germany	Marianne Darbi
October 2017	

Acknowledgements

This book is an outcome of IOER activities around the topic "Biodiversity Offsets: Towards a European No Net Loss Strategy for the Protection of Biodiversity". The book has been initiated by the Leibniz Institute of Ecological Urban and Regional Development Germany (IOER) and is implemented and compiled by editors from IOER, the Institute for European Environmental Policy (IEEP), BIOTOPE and the Helmholtz Centre for Environmental Research (UFZ). The activities on this publication have been supported by the German Federal Agency for Nature Conservation. The preparation of the book benefited from support from the European Commission Seventh Framework Programme under Grant Agreement No. FP7-ENV-2012-308393-2 (OPERAs).

We would kindly like to thank all the authors for providing fruitful and inspiring contributions illustrating already so many activities on biodiversity offsets around Europe. We also would like to thank Derek Henderson for language polishing. We particularly would like to say thank you to Sabine Witschas, Dipl.-Ing. Cartography at IOER, and her team comprising Margitta Wahl, Natalija Leutert, Jenny Herbrich and others for improving the figures and for helping us compile and review the text. Special thanks to our Director of IOER, Prof. Bernhard Müller, who did a lot to make this book a reality. The organisation and the cooperative effort with our publisher, Springer, was carried out in a notably pleasant atmosphere.

"Where now cities stand, a meadow will be"
"Wo jetzt noch Städte stehn, wird eine Wiese sein"
Andreas Gryphius ca. 1630
Excerpt from the poem "All is Vanity" translated by Scott Horton

Disclaimer:

This book is a review by independent researchers and does not represent the views of the Commission.

Contents

List of Contributors

Juliane Albrecht Leibniz Institute of Ecological Urban and Regional Development, Dresden, Germany

Martina Artmann Leibniz Institute of Ecological Urban and Regional Development, Dresden, Germany

Vytaute Bacianskaite Deutsche Bundesstiftung Umwelt 2015, Vilnius, Lithuania

Jonathan Baker London, UK

Adeline Bas AMURE, Université de Brest, IFREMER, CNRS, IUEM / AMURE Unité d'Economie Maritime, IFREMER, Université de Brest, CNRS, IUEM, Plouzané, France

Lucie Bezombes EDF/IRSTEA/National Museum of National History, Grenoble, France

Charles-Hubert Born University of Louvain, Louvain-la-Neuve, Belgium

Adéla Boučková Charles University, Prague, Czech Republic

Coralie Calvet CEE-M, CNRS, Université de Montpellier, INRA, SupAgro, Université Paul Valéry Montpellier, Montpellier, France

Marianne Darbi Helmholtz Centre for Environmental Research, Leipzig, Germany

Valérie Dupont University of Louvain, Louvain-la-Neuve, Belgium

David Álvarez Garcia Ecoacsa Reserva de Biodiversidad, Madrid, Spain

Stéphanie Gaucherand UR EMGR, Université Grenoble Alpes, Irstea, St-Martin-d'Hères, France

Amrei von Hase Forest Trends, Cape Town, South Africa

Julien Hay AMURE, Université de Brest, IFREMER, CNRS, IUEM / AMURE Unité d'Economie Maritime, IFREMER, Université de Brest, CNRS, IUEM, Plouzané, France

Matthias Herbert Federal Agency for Nature Conservation, Leipzig, Germany

Céline Jacob AMURE, Université de Brest, IFREMER, CNRS, IUEM / AMURE Unité d'Economie Maritime, IFREMER, Université de Brest, CNRS, IUEM, Plouzané, France

Kerry ten Kate Forest Trends, Hampshire, UK

Charlène Kermagoret Université du Québec en Outaouais, Gatineau, Canada

Harold Levrel CIRED, AgroParisTech, Nogent-sur-Marne, France

Patrick Maguire Forest Trends, Washington DC, USA

Aurore Malapert BIOTOPE Chine – Feilan, Beijing, China

Anja May Berlin, Germany

Liza Papadopoulou Fairfax, VA, USA

Sylvain Pioch CEFE, Université Paul-Valéry Montpellier 3, CNRS, Université de Montpellier, EPHE, Montpellier, France

Fabien Quétier BIOTOPE, Mèze, France

Matt Rayment Consultant, London, UK

Marie-Eve Reinert Karlsruhe, Germany

Pierre Scemama AMURE, Université de Brest, IFREMER, CNRS, IUEM / AMURE Unité d'Economie Maritime, IFREMER, Université de Brest, CNRS, IUEM, Plouzané, France

Hendrik Schoukens Ghent University, Ghent, Belgium

Jochen Schumacher Tübingen, Germany

William Sheate Collingwood Environmental Planning, London, UK

Martin Szaramowicz Compensation Agency Brandenburg, Brandenburg an der Havel, Germany

Astrid J.A. van Teeffelen Faculty of Science, Environmental Geography Group, Vrije Universiteit, Amsterdam, The Netherlands

Graham Tucker Institute for European Environmental Policy, London, UK

Anne-Charlotte Vaissière Ecologie Systématique Evolution, CNRS, AgroParis Tech, Université Paris-Sud, Université Paris-Saclay, Orsay, France

Wolfgang Wende Leibniz Institute of Ecological Urban and Regional Development and TU Dresden, Dresden, Germany

National University of Singapore, Singapore, Singapore

Introduction: Biodiversity Offsets—The European Perspective on No Net Loss of Biodiversity and Ecosystem Services

Wolfgang Wende, Graham Tucker, Fabien Quétier, Matt Rayment, and Marianne Darbi

One of today's great challenges is to safeguard biodiversity for future generations. An increasing number of countries and economic sectors are adopting strategies to slow and stop biodiversity loss. In this book, we examine existing approaches to achieve no net loss of biodiversity in selected EU countries. Adopting an explicitly European perspective, our focus is on various options to compensate negative impacts on nature, ecosystem services and biodiversity. Although Europe is crisscrossed by a network of protected areas such as Natura 2000 sites, and the EU has a strict legal framework for offsetting within Natura 2000, this is not enough to conserve the continent's biodiversity and the continued flow of benefits from nature to people. Europe is a fine-grained patchwork of natural and modified habitats which form varied and highly valued cultural landscapes. These provide vital ecosystem services as well as habitats for animals and plants, resulting in a high level of

W. Wende (✉)
Leibniz Institute of Ecological Urban and Regional Development and TU Dresden, Dresden, Germany

National University of Singapore, Singapore, Singapore
e-mail: w.wende@ioer.de

G. Tucker
Institute for European Environmental Policy, London, UK
e-mail: GTucker@ieep.eu

F. Quétier
BIOTOPE, Mèze, France
e-mail: fquetier@biotope.fr

M. Rayment
Consultant, London, UK
e-mail: matt.rayment@outlook.com

M. Darbi
Helmholtz Centre for Environmental Research, Leipzig, Germany
e-mail: marianne.darbi@ufz.de

W. Wende et al. (eds.), *Biodiversity Offsets*,
https://doi.org/10.1007/978-3-319-72581-9_1

biodiversity outside of dedicated nature reserves. This book emphasizes the values of nature and biodiversity in landscapes shaped by human activity.

The European Union has introduced a strategy to halt the loss of biodiversity by the year 2020. One of the actions is an initiative to ensure no net loss of ecosystems and their services. Recognizing the need to protect wider biodiversity and the benefits it provides, also beyond the Natura 2000 network of protected areas, the recently adopted EU action plan on nature, people and the economy (2017–2019) launched an initiative to integrate ecosystems and their services into planning and decision-making. A guidance document, to be published in 2018, will also provide an overview of approaches to apply mitigation hierarchy in relation to impacts on ecosystems and their services. This EU initiative underlines the urgent need for sharing experiences and increasing knowledge on policies, legislation and standards to stimulate and regulate the offsetting of biodiversity and ecosystem services.

Of course, Europe is not starting from scratch. There already exist numerous national models for compensating impacts on biodiversity and other ecosystem values. Later chapters provide an overview of these models, revealing common features as well as differences between countries. First, however, an introductory section examines the requirements of the EU's Biodiversity Strategy 2020 (see chapter "Introduction of a European Strategy on No Net Loss of Biodiversity") as well as presenting the global and principles-based framework of the Business and Biodiversity Offsets Programme (see chapter "Principles of the Business and Biodiversity Offsets Programme"). In chapter "Introduction of a European Strategy on No Net Loss of Biodiversity" we also consider the central points in the European debate on no net loss of biodiversity as they relate to the wider global discussion. Existing national approaches and offset models are then presented in detail. Specifically, the book presents current policies and practice in Austria (see chapter "Austria"), Belgium (see chapter "Belgium"), Czech Republic (see chapter "The Czech Republic"), France (see chapter "France"), Germany (see chapter "Germany"), Lithuania (see chapter "Lithuania"), Spain (see chapter "Spain"), The Netherlands (see chapter "The Netherlands") and the United Kingdom (see chapter "United Kingdom").

Although only a small selection of European countries is included here, the editors believe that those chosen—while they are by no means comprehensive in capturing experience across the EU in its entirety—demonstrate a valuable range of insights into the potential role of offsets in different contexts. Further research to extend the analysis to the EU as a whole, and to other European countries, would be valuable. In chapter "Other EU Member States" we attempt to partially fill this gap by giving a brief overview of alternative approaches within Europe such as offsetting concepts adopted in Denmark, Finland, Sweden, Switzerland and Hungary. Literature references are also provided for those wishing to read more about these concepts. In chapter "Conclusions: Lessons from Biodiversity Offsetting Experiences in Europe" we draw some conclusions and give an outlook on additional EU activities.

Each country chapter follows the same basic structure. First, a summary is given of the national legal framework for no net loss and biodiversity offsetting. This is followed by a presentation of the theoretical approach and adopted methods, including, for example, the ways in which diverse metrics are applied in each country. After this overview of the adopted methodology in each country, examples of biodiversity offsets are given. Finally, each chapter closes with a discussion of the national approach to habitat "banking", provided that the country has adopted such a banking model. In some cases, offsets and "banking" remain innovative and case studies couldn't be found for all countries. The included examples, however, illustrate the fact that offsets succeed *only* if they are well regulated, and follow appropriate technical requirements, and are not simply viewed as a 'licence to trash'. If we are serious about implementing such a strategy and if we clearly obey the corresponding principles (notably the BBOP principles presented in chapter "Principles of the Business and Biodiversity Offsets Programme"), then the examples in this book confirm that offsets can help to prevent further net loss of biodiversity and ecosystem services in years to come.

Efforts to halt and reverse the loss of biodiversity can greatly benefit from building on existing national approaches. We hope, therefore, that our book will find a wide readership, and contribute to achieving the EU biodiversity headline target of halting the loss of biodiversity and ecosystem services by 2020.

Introduction of a European Strategy on No Net Loss of Biodiversity

Wolfgang Wende, Lucie Bezombes, and Marie-Eve Reinert

A broad and intensive discussion has been ongoing within the EU and at the level of its Member States on how to achieve no net loss of biodiversity by the year 2020. This European discussion is embedded in a wider and at times controversial global debate on whether and how to implement new tools such as biodiversity offsets for managing impacts on biodiversity from economic developments beyond the already existing requirements of Natura 2000 conservation and offsetting (ten Kate et al. 2004; Gordon et al. 2011, 2015; The Business and Biodiversity Offsets Programme/BBOP 2012a; Gardner et al. 2013; Quétier et al. 2014; Maron 2015; Maron et al. 2015a; Vaissière and Levrel 2015; Darbi in preparation).

The EU Biodiversity Strategy to 2020, published in 2011, aims "to halt the loss of biodiversity and the degradation of ecosystem services in the EU by 2020, and to restore them in so far as feasible". Under the strategy's Target 2 ("restore and enhance ecosystems and their services") Action 7 seeks to "ensure no net loss of biodiversity and ecosystem services". Action 7b specifies that "the Commission will carry out further work with a view to propose an initiative to ensure there is no net

W. Wende (✉)
Leibniz Institute of Ecological Urban and Regional Development and TU Dresden, Dresden, Germany

National University of Singapore, Singapore, Singapore
e-mail: w.wende@ioer.de

L. Bezombes
EDF/IRSTEA/National Museum of National History, Grenoble, France
e-mail: lucie.bezombes@irstea.fr

M.-E. Reinert
Karlsruhe, Germany
e-mail: marieeve.reinert@gmail.com

W. Wende et al. (eds.), *Biodiversity Offsets*,
https://doi.org/10.1007/978-3-319-72581-9_2

loss of ecosystems and their services by 2015 (e.g. through compensation or offsetting schemes)" (see European Commission website[1] and Tucker et al. 2014, 2016). This commitment is currently being taken forward within a wider initiative to integrate ecosystems and their services into decision-making, as part of the Action plan for nature, people and the economy (2017–2019). In the following chapters, we examine activities which could be highly relevant to this initiative and discuss existing concepts, legislation and practices in several EU member states.

Options to achieve no net loss of biodiversity have been intensively discussed by the EU Commission and within individual EU member states (see Tucker et al. 2014 and follow-up studies). Much effort has already been expended on building a shared understanding of the issues and on reframing of existing environmental impact assessment and permitting processes in terms of delivering measurable biodiversity outcomes, such as no net loss of biodiversity. This requires that developers, project proponents and permitting authorities pay special attention to minimizing the impacts of their projects, and implement suitable offsets to compensate for those impacts with equivalent or comparable biodiversity gains elsewhere.

A key point in this discussion is that biodiversity offsetting is to be understood as a tool to achieve no net loss of biodiversity in wider contexts than the legislative requirements under the EU Habitats Directive. Thus, it is to be applied beyond the scope of strictly protected nature conservation areas and to species and habitats that are not in the focus of the Habitats and Birds Directives. Many stakeholders have expressed a wish that biodiversity offsets be applied first and foremost to manage impacts on 'normal' landscapes, not areas of outstanding conservation value, including Natura 2000 sites protected under the Birds and Habitats Directives. The latter, however, still suffer from direct or indirect negative impacts. There is scope to strengthen the implementation of the EU Nature Directives with a focus on avoiding such negative impacts and, for exceptional cases of unavoidable damage, to achieve no net loss, or even net gains for protected habitats and species (Milieu et al. 2016).

There are many other issues raised by the no net loss (NNL) objective, and before describing national approaches and experiences (in the following chapters), we first present below some key scientific and practical issues through a descriptions of key terms: 'mitigation hierarchy', 'biodiversity offsets/compensation', 'land use change, project orientation', 'biodiversity and ecosystem services', 'metrics', 'ecological equivalency', 'additionality' and 'habitat banking'. This draws on a terminology that was developed by an EU level expert working group on NNL (Dickie et al. 2013).

The terminology presented here should not be understood as a rigorous *definition* of terms, as there are variations on their technical interpretations, i.e. there can be divergences in the use of terminology depending on national or cultural contexts. Therefore, the descriptions of key terms given below are merely an aid to comprehend the general approach of no net loss of biodiversity and biodiversity offsetting.

---

[1] http://ec.europa.eu/environment/nature/biodiversity/nnl/index_en.htm

1 Mitigation Hierarchy

According to widely accepted international principles, biodiversity offsetting should follow a mitigation hierarchy (MH) when attempting to achieve no net loss of biodiversity. Within the EU's glossary on no net loss, mitigation hierarchy is defined as follows:

"Biodiversity offset/compensation schemes usually follow a three step mitigation hierarchy of:

- Avoid or prevent negative impacts on the environment in general and biodiversity in particular;
- Minimise and rehabilitate on-site effects of development if impacts cannot be avoided; and
- Offset/compensation measures that are undertaken as a last resort (on or off-site) for the residual adverse impacts." (see Dickie et al. 2013)

Nearly all descriptions of the mitigation hierarchy include the three basic steps of avoidance, minimization and then offsets/compensation. Although this is applied as a general principle, the terminology varies considerably from one country to another (e.g. Jacob et al. 2016a, b) and some definitions break the MH into more steps: both BBOP and the International Finance Corporation include a step titled 'restoration/ rehabilitation' before offsets and distinguish between offsets and compensation as a last resort. A key principle is that offsets cannot provide a justification for proceeding with projects for which the residual impacts on biodiversity are unacceptable (e.g., loss of significant proportion of population of a protected species). This means that the avoidance options have to be considered seriously in harmful cases (Dickie et al. 2013) and that non-offsetable impacts should be defined (BBOP 2012a).

This illustrates the priority of a reasonable avoidance and minimization before impacts are remedied by means of biodiversity offset measures. Under the mitigation hierarchy, therefore, we must first strive for an actual avoidance and a natural rehabilitation, compensation/offset before turning to offset payments as a last resort (in some countries this last alternative is not even allowed). Offset payments should only be allowed when actual physical offsets are not realizable.

2 Biodiversity Offsets/Compensation

The BBOP defines biodiversity offsets as "measurable conservation outcomes resulting from actions designed to compensate for significant residual adverse biodiversity impacts arising from project development after appropriate prevention and mitigation measures have been taken. The goal of biodiversity offsets is to achieve no net loss and preferably a net gain of biodiversity on the ground with respect to species composition, habitat structure and ecosystem function and people's use and cultural values associated with biodiversity" (Business and Biodiversity Offsets Programme 2007 and chapter "Principles of the Business and Biodiversity Offsets Programme" in this book).

More generally, and in contrast to the BBOP definition of biodiversity offsets, compensation is a recompense for some loss or service which constitutes an equivalent replacement for the lack or variation of something else. It involves that something (such as money) is given or received as payment or reparation (as for a service or loss or injury). Specifically, in terms of biodiversity, compensation involves measures to recompense, make good or pay damages for the loss of biodiversity caused by a project or an environmental accident. However, it is important to note that the 'compensation measures' that are required for unavoidable impacts on Natura 2000 sites under the Habitats Directive (Art. 6, 4) are analogues to offsets. Similarly, in some languages and the concepts of EU member states (reflected in some sections of this book) 'compensation' is used synonymously with 'offset'. However, the BBOP makes a clear distinction between compensation and biodiversity offsets (see Dickie et al. 2013).

Experts have referred to the BBOP definition as appropriate to use in the context of EU level discussions on no net loss: "BBOP's definition that draws a distinction between more general compensation that could include indirect measures such as awareness campaigns and financial payments, and offset mechanisms that have explicit NNL or net gain goals would seem appropriate" (Dickie et al. 2013).

3 Land Use Change, Project Orientation

The term *land use change* should be applied to all kinds of changes in land use resulting from plane, linear or punctual projects. The term is thus project-related, and any land use change significantly impacting biodiversity and ecosystem services should be subject to the no net loss principle. Many scientific and policy papers maintain that the application of no net loss principle should cover any impact on the environment or biodiversity with no exceptions made for agricultural, fishery and forestry usage (Aiama et al. 2015). All impacts should be considered insofar as they are significant in regard to nature and landscapes. Furthermore, no exceptions should be made for supposedly 'environmentally-friendly' projects, e.g. the production of regenerative energy such as wind turbines, as these can also have a significant impact on biodiversity (cf. Gartman et al. 2016).

4 Biodiversity and Ecosystem Services

While the approach of some member states is to focus purely on biodiversity [i.e. in majority biodiversity components subjects to protection such as plants, animals species and habitats considered particularly important for conservation mainly because they present a significant risk of extinction in the medium term (EEC) 1992, 2009], others consider biodiversity as well as ecosystem functions or services such as groundwater recharge services or other soil services (see Jacob et al. 2016a, b and Grunewald and Bastian 2015 for the incorporation of ecosystem services). A few even adopt the approach of considering landscape aesthetic qualities to achieve no net loss.

5 Metrics

The EU level working group on NNL, offered a working definition of the term "metrics" as follows (Dickie et al. 2013): Metrics are a "set of unitary measurements of biodiversity lost, gained or exchanged. This varies from very basic measures such as area analysis, to sophisticated quantitative indices of multiple biodiversity components which may be variously weighted. These metrics are used in order to compare losses at the damaged site and gains at the compensation site and provide decision support to establish equivalence."

The aim of metrics is to support the decision-making process. Although metrics should be based on scientific and empirical evidence, they do not always assess the ecological state or functioning of a site or target biodiversity in a purely scientific manner (Bezombes et al. 2017). Different approaches are developed in practice, involving rule-based models, more verbal descriptive or calculation models that adopt certain algorithms, depending on the resources available and the people involved. All of these approaches pursue the same aim of estimating the importance or value of some ecological feature that can be compared in a before-after biodiversity status assessment (see Dickie et al. 2013).

ICF International and IEEP (2014, 6) describe the aim of metrics as follows: "Offsets aim to ensure no net loss (or a net gain) of biodiversity and this should apply to all components of biodiversity importance that are significantly impacted. Therefore, to ensure this objective is achieved it is necessary to measure biodiversity losses from impacts (biodiversity debits) and the gains from offsets (biodiversity credits) in a practical and transparent way so that their equivalency can be compared."

6 Ecological Equivalence

For the BBOP "the term is synonymous with the concept of 'like for like' and refers to areas with highly comparable biodiversity components. This similarity can be observed in terms of species diversity, functional diversity and composition, ecological integrity or condition, landscape context (e.g., connectivity, landscape position, adjacent land uses or condition, patch size, etc.), and ecosystem services (including people's use and cultural values)" (BBOP 2012b).

In a general way, ecological equivalence is reached between biodiversity losses and gains when they are considered equals in terms of the metrics used to assess them (see the metrics description). Implementing offset measures equivalent to impact is one of the key conditions for offsets to achieve no net loss. In this purpose, assessing equivalence should take into account some essential considerations (Bezombes et al. 2017). First, careful attention should be given to the metrics chosen as surrogate for target biodiversity or ecosystem services. Also, equivalence assessment should integrate spatial considerations as the integration of impacted and compensatory sites landscape. According to the BBOP (2012c) "a biodiversity offset should be designed and implemented in a landscape context to achieve the expected measurable conservation outcomes". As there are time lags between the event when impacts on biodiversity occur and the moment when offset measures become fully effective, equivalence assessment should also consider this delay in order to minimize interim losses. Finally, it is of great importance to take into account the lack of confirmed knowledge and hindsight when assessing equivalence and particularly the risk of failure when implementing offset measures (Curran et al. 2013).

7 Additionality

Biodiversity offsets should contribute to a new and additional outcome in nature conservation. The simple preservation of already natural valuable sites cannot lead to no net loss, and thus does not fully follow the principle of additionality. BBOP gives a quite clear definition of additionality: "A biodiversity offset should achieve conservation outcomes above and beyond results that would have occurred if the offset had not taken place."

McKenney and Kiesecker (2010) provide a similar explanation of additionality, referring to the need for a compensation or offset measure to provide a new contribution to conservation in addition to any existing values, i.e. the conservation outcomes it delivers would not have otherwise occurred (see also Dickie et al. 2013). This is directly related to the question of whether or not to implement offset measures in already protected nature conservation areas. On the one hand, this can make sense as long as it actually creates a value added to the conservation results that would have occurred anyway due to the conservation management. On the other hand, and over the long-term, this approach can lead to an ever increasing division between 'valuable' landscapes where all nature conservation efforts are to be concentrated as against more impacted landscapes where biodiversity is undermined. Hence, the use of protected areas for implementing offsets is an option that should only be pursued if there is no other feasible way of implementing on-site and off-site measures within the ordinary landscape *and* if a real additionality is created (see also Maron et al. 2015b). Designing offset measures leading to additional outcomes implies to choose a baseline upon which to assess and measure additionality (see Dickie et al. 2013). Examples in the following chapters will help illustrate how such a baseline can be defined and fixed in practice.

8 Habitat Banking

A study elaborated for the EU's DG Environment describes habitat banking as "a market where the credits from actions with beneficial biodiversity outcomes can be purchased to offset the debit from environmental damage. Credits can be produced in advance of, and without ex-ante links to, the debits they compensate for, and stored over time" (Dickie et al. 2013) (Fig. 1).

Froger et al. (2015) state that the considerable lack of clarity regarding biodiversity and habitat banks as well as their institutional features and some other aspects have led to a number of misconceptions. The purpose of their article (Froger et al. 2015) is to clarify the concept of biodiversity and/or habitat banking and its effects. In some languages and member states' concepts (reflected in this book), the terms 'conservation banking', 'mitigation banking', 'bio banking' or even 'eco account' are used synonymously with 'habitat banking'. However 'habitat or conservation banking' seems to have established itself as the preferred terminology within the EU-wide discussion (see Dickie et al. 2013). These banks are considerably less developed in Europe compared to the USA (Committee on Mitigating Wetland Losses 2002 and Kozich and Halvorsen 2012) where the concept was born, but

Fig. 1 Alpine grassland in Combe Madame (photo: Bezombes)

Fig. 2 Restauration of black goose habitat already implemented partially in this sector of Combe Madame (photo: Bezombes)

increasing experiments start to emerge in EU member states (e.g. France, Spain and others; see also van Teeffelen et al. 2014) and some countries already largely use habitat banking for biodiversity offset (Germany). The pictures show an experimentation of a French habitat bank in the Alps at Combe Madame conducted by the association IBCM (Combe Madame Biodiversity Initiative). Different types of offset measures are going to be implemented which will constitute different kind of biodiversity units: restauration of black goose habitat (already implemented partially in the sector on Fig. 2) and diversification of woodland by opening areas, edge creation and senescent trees conservation (expected to be implemented in the sector on Fig. 3).

Fig. 3 Diversification of woodland by opening areas, edge creation and senescent trees conservation expected to be implemented in this sector of Combe Madame (photo: Bezombes)

References

Aiama D, Edwards S, Bos G, Ekstrom J, Krueger L, Quétier F, Savy C, Semroc B, Sneary M, Bennun L (2015) No net loss and net positive impact approaches for biodiversity: exploring the potential application of these approaches in the commercial agriculture and forestry sectors. International Union for Conservation of Nature. https://portals.iucn.org/library/sites/library/files/documents/2015-003.pdf. Last Access 7 Feb 2017

Bezombes L, Gaucherand S, Kerbiriou C, Reinert ME, Spiegelberger T (2017) Ecological equivalence assessment methods: what trade-offs between operationality, scientific basis and comprehensiveness? Environ Manag 60(2):216–230. https://doi.org/10.1007/s00267-017-0877-5

Business and Biodiversity Offsets Programme (2007) BBOP principles on biodiversity offsets. http://bbop.forest-trends.org/documents/files/bbop_principles.pdf. Last Access 30 Jan 2017

Business and Biodiversity Offsets Programme (BBOP) (2012a) Resource paper: limits to what can be offset. Business and Biodiversity Offsets Programme, Washington, DC

Business and Biodiversity Offsets Programme (BBOP) (2012b) Glossary, 2nd updated edn. BBOP, Washington, DC

Business and Biodiversity Offsets Programme (BBOP) (2012c) Resource paper: no net loss and loss-gain calculations in biodiversity offsets. BBOP, Washington, DC

Committee on Mitigating Wetland Losses (2002) Compensating for wetland losses under the clean water act. National Academy, Washington, DC

Curran M, Hellweg S, Beck J (2013) Is there any empirical support for biodiversity offset policy? Ecol Appl 24:617–632

Darbi M (in preparation) Biodiversity offsets between regulation and voluntary commitment. Springer Publishing House. Dissertation, Technische Universität Dresden, Germany

Dickie I, Mesquite S, Morandeau D, Stoeckel M-E, Weaver J, Wende W (2013) Glossary working group on no net loss of ecosystems and their services. European Commission, DG Environment. http://ec.europa.eu/environment/nature/biodiversity/nnl/pdf/NNL_Glossary.pdf. Last Access 27 Jan 2017

EEC (1992) Council directive 92/43/EEC of 21 May 1992 on the conservation of natural habitats and of wild fauna and flora. Off J Eur Union 206:7–50

EEC (2009) Directive 2009/147/EC of the European parliament and of the council of 30 November 2009 on the conservation of wild birds on the conservation of wild birds (codified version). Off J L20:7–25

Froger G, Ménard S, Méral P (2015) Towards a comparative and critical analysis of biodiversity banks. Ecosyst Serv 15:152–161

Gardner T-A, von Hase A, Brownlie S, Ekstrom J-M-M, Pilgrim J-D, Savy C-D, Stephens R-T-T, Treweek J, Ussher G-T, Ward G, Ten Kate K (2013) Biodiversity offsets and the challenge of achieving no net loss. Conserv Biol 27(6):1254–1264

Gartman V, Bulling L, Dahmen M, Geißler G, Köppel J (2016) Mitigation measures for wildlife in wind energy development, consolidating the state of knowledge—part 1: planning and siting, construction. J Environ Assess Policy Manag 18(3):1650013. https://doi.org/10.1142/S1464333216500137

Gordon A, Langford W-T, Todd J-A, White M-D, Mullerworth D-W, Bekessy S-A (2011) Assessing the impacts of biodiversity offset policies. Environ Model Softw 2(6):1481–1488

Gordon A, Bull J-W, Wilcox C, Maron M (2015) Forum – perverse incentives risk undermining biodiversity offset policies. J Appl Ecol 52(2):532–537. https://doi.org/10.1111/1365-2664.12398

Grunewald K, Bastian O (eds) (2015) Ecosystem services – concept, methods and case studies. Springer, Berlin

ICF International, IEEP (2014) Study on specific design elements of biodiversity offsets: biodiversity metrics and mechanisms for securing long term conservation benefits. DG ENVIRONMENT: ENV.B.2/ETU/2013/0060r. Final Report. http://ec.europa.eu/environment/nature/biodiversity/nnl/pdf/Biodiversity%20offsets%20metrics%20and%20mechanisms.pdf. Last Access 27 Jan 2017

Jacob C, Pioch S, Thorin S (2016a) The effectiveness of the mitigation hierarchy in environmental impact studies on marine ecosystems: a case study in France. Environ Impact Assess Rev 60:83–98

Jacob C, Vaissiere A-C, Bas A, Calvet C (2016b) Investigating the inclusion of ecosystem services in biodiversity offsetting. Ecosyst Serv 21:92–102

Kozich A-T, Halvorsen K-E (2012) Compliance with wetland mitigation standards in the Upper Peninsula of Michigan, USA. Environ Manag 50:97–105

Maron M (2015) Stop misuse of biodiversity offsets. Nature 523:401–403

Maron M, Bull J-W, Evans M-C, Gordon A (2015a) Locking in loss: baselines of decline in Australian biodiversity offset policies. Biol Conserv 192:504–512

Maron M, Gordon A, Mackey B-G, Possingham H-P, Watson J-E-M (2015b) Interactions between biodiversity offsets and protected area commitments: avoiding perverse outcomes. Conserv Lett 9(5):384–389. https://doi.org/10.1111/conl.12222

McKenney BA, Kiesecker JM (2010) Policy development for biodiversity offsets: a review of offset frameworks. Environ Manag 45(1):165–176

Milieu, IEEP, ICF (2016) Evaluation study to support the fitness check of the birds and habitats directives. Milieu, Institute for European Environmental Policy and the ICF International, Brussels

Quétier F, Regnery B, Levrel H (2014) No net loss of biodiversity or paper offsets? A critical review of the French no net loss policy. Environ Sci Pol 38:120–131

ten Kate K, Bishop J, Bayon R (2004) Biodiversity offsets: views, experience, and the business case. IUCN and Insight Investment, Gland

Tucker G-M, Allen B, Conway M, Dickie I, Hart K, Rayment M, Schulp C, van Teeffelen A (2014) Policy options for an EU no net loss initiative. Report to the European Commission. Institute for European Environmental Policy, London. http://ec.europa.eu/environment/nature/biodiversity/nnl/pdf/Policy%20Options.pdf. Last Access 23 Jun 2016

Tucker G-M, Dickie I, McNeil D, Rayment M, ten Brink P, Underwood E (2016) Supporting the elaboration of the impact assessment for a future EU initiative on no net loss of biodiversity and ecosystem services. Report to the European Commission. Institute for European Environmental Policy, London

Vaissière A-C, Levrel H (2015) Biodiversity offset markets: what are they really? An empirical approach to wetland mitigation banking. Ecol Econ 110:81–88

van Teeffelen A-J-A, Opdam P, Wätzold F, Hartig F, Johst K, Drechsler M, Vos C-C, Wissel S, Quétier F (2014) Ecological and economic conditions and associated institutional challenges for conservation banking in dynamic landscapes. Landsc Urban Plan 130:64–72

Principles of the Business and Biodiversity Offsets Programme

Kerry ten Kate, Amrei von Hase, and Patrick Maguire

1 The Aims and Origins of the Business and Biodiversity Offsets Programme (BBOP)

In 2004, Forest Trends established the Business and Biodiversity Offsets Programme (BBOP) to bring together a large group of organisations to challenge the historical assumption that the social and economic benefits of development projects must inevitably result in a net loss of biodiversity. At the time, companies were beginning to acknowledge that the trade-off between economic growth and environmental outcomes was increasingly unacceptable to investors and civil society. Governments were looking for practical ways to reconcile their sustainable development targets with biodiversity conservation. Financial institutions wanted to find ways to safeguard their investments against social and environmental risks. Indigenous peoples and local communities wanted to ensure that new projects were developed with their prior and informed consent and reflected their needs and priorities. The conservation community and scientists aimed to improve the manner in which losses and gains of biodiversity and ecosystem services were measured, managed and monitored and to ensure that conservation priorities and land-use planning were based on sound science. All of them faced challenges in making progress with these goals. The terminology for core concepts such as 'mitigation', 'compensation' and 'offsets'

K. ten Kate (✉)
Forest Trends, Hampshire, UK
e-mail: ktenkate@forest-trends.org

A. von Hase
Forest Trends, Cape Town, South Africa
e-mail: avonhase@forest-trends.org

P. Maguire
Forest Trends, Washington DC, USA
e-mail: pmaguire@forest-trends.org

W. Wende et al. (eds.), *Biodiversity Offsets*,
https://doi.org/10.1007/978-3-319-72581-9_3

varied from country to country and group to group, leading to confusion and misunderstanding during discussions; guidelines, methodologies and standards were lacking; proposals for improved approaches hadn't been tested and demonstrated at pilot sites; and government policies and financial investment conditions did not necessarily encourage best practice. With this in mind, 40 representatives from companies, governments, non-governmental organisations and financial institutions joined BBOP; a group that has now grown to over 80 members, with a Secretariat provided by Forest Trends and WCS. The plan was to develop and test the principles, standards and methods needed to demonstrate no net loss of biodiversity in the context of development projects.

2 An Approach Based on Principles

Early in the group's discussions, the BBOP members decided to take a principles-based approach to defining best practice on mitigation, offsets and no net loss. Rather than endeavouring to prescribe very detailed guidelines to be applied to every scenario, from a trans-continental pipeline to a marine oil and gas development to a micro-scale tourism lodge, members agreed that best practice should be established by defining a set of principles that set a high standard on how to proceed but that are flexible enough to apply in very varied circumstances. Another advantage of defining principles is that they can be consistent with a range of regulatory or voluntary approaches that exist in different jurisdictions. Box 1 contains the set of principles supported by the members of the BBOP Advisory Group. Members developed and agreed these by consensus based on practical experience with projects and drawing on common elements of frameworks for best practice in law, policy and corporate commitments from around the world. Approaches to mitigation that follow these principles should achieve the best outcomes for biodiversity and manage the risks associated with biodiversity offsets, since they stress the primacy of 'avoidance' and the correct position of offsets as a last step, guide key aspects such as exchange rules, metrics, additionality and landscape-level planning, require evidence of permanence, and provide the basis for a rights-based approach, participation and transparency.

Box 1 Principles on Biodiversity Offsets Supported by the Members of the BBOP Advisory Group

Biodiversity offsets are measurable conservation outcomes resulting from actions designed to compensate for significant residual adverse biodiversity impacts arising from project development[1] after appropriate prevention and

(continued)

[1]While biodiversity offsets are defined here in terms of specific development projects (such as a road or a mine), they could also be used to compensate for the broader effects of programmes and plans.

Box 1 (continued)

mitigation measures have been taken. The goal of biodiversity offsets is to achieve no net loss and preferably a net gain of biodiversity on the ground with respect to species composition, habitat structure, ecosystem function and people's use and cultural values associated with biodiversity.

These principles establish a framework for designing and implementing biodiversity offsets and verifying their success. Biodiversity offsets should be designed to comply with all relevant national and international law, and planned and implemented in accordance with the Convention on Biological Diversity and its ecosystem approach, as articulated in National Biodiversity Strategies and Action Plans.

1. *Adherence to the mitigation hierarchy*: A biodiversity offset is a commitment to compensate for significant residual adverse impacts on biodiversity identified after appropriate avoidance, minimization and on-site rehabilitation measures have been taken according to the mitigation hierarchy.
2. *Limits to what can be offset*: There are situations where residual impacts cannot be fully compensated for by a biodiversity offset because of the irreplaceability or vulnerability of the biodiversity affected.
3. *Landscape context*: A biodiversity offset should be designed and implemented in a landscape context to achieve the expected measurable conservation outcomes taking into account available information on the full range of biological, social and cultural values of biodiversity and supporting an ecosystem approach.
4. *No net loss*: A biodiversity offset should be designed and implemented to achieve in situ, measurable conservation outcomes that can reasonably be expected to result in no net loss and preferably a net gain of biodiversity.
5. *Additional conservation outcomes*: A biodiversity offset should achieve conservation outcomes above and beyond results that would have occurred if the offset had not taken place. Offset design and implementation should avoid displacing activities harmful to biodiversity to other locations.
6. *Stakeholder participation*: In areas affected by the project and by the biodiversity offset, the effective participation of stakeholders should be ensured in decision-making about biodiversity offsets, including their evaluation, selection, design, and implementation and monitoring.
7. *Equity*: A biodiversity offset should be designed and implemented in an equitable manner, which means the sharing among stakeholders of the rights and responsibilities, risks and rewards associated with a project and offset in a fair and balanced way, respecting legal and customary arrangements. Special consideration should be given to respecting both

(continued)

Box 1 (continued)

internationally and nationally recognized rights of indigenous peoples and local communities.

8. *Long-term outcomes*: The design and implementation of a biodiversity offset should be based on an adaptive management approach, incorporating monitoring and evaluation, with the objective of securing outcomes that last at least as long as the project's impacts and preferably in perpetuity.
9. *Transparency*: The design and implementation of a biodiversity offset, and communication of its results to the public, should be undertaken in a transparent and timely manner.
10. *Science and traditional knowledge*: The design and implementation of a biodiversity offset should be a documented process informed by sound science, including an appropriate consideration of traditional knowledge.

3 The Standard on Biodiversity Offsets

Principles such as those developed by BBOP have the advantage of being broadly applicable, flexible and brief, but they are aspirational in nature. How can one tell whether the principles such as 'adhere to the mitigation hierarchy' have been met? How can one know whether a given project has been designed and is being implemented in such a way that it is on track to result in no net loss or a net gain of biodiversity? In order to answer these questions, the BBOP members have developed a Standard (BBOP 2012a). This is presented as a hierarchy of Principles, Criteria and Indicators (PCI): an architecture similar to that used in a number of other social and environmental standards. 'Principles' are interpreted as the fundamental statements about a desired outcome. 'Criteria' are the conditions that need to be met in order to comply with a Principle. 'Indicators' are the measurable states which allow the assessment of whether or not a particular Criterion has been met. Taken together, a standard of this kind can enable a developer, its partners or independent third party verifiers to establish whether the plans for mitigation (including offsets) have reached or are on course to achieve 'No Net Loss'.

The Standard was designed through a collaborative process following the ISEAL Code of Good Practice for Setting Social and Environmental standards, several rounds of public consultations, and testing at pilot sites. The BBOP members originally had two purposes in mind for the Standard: to enable auditors and assessors to determine whether an existing set of mitigation measures (including biodiversity offsets) comply with the Principles, Criteria and Indicators; and to guide the design and implementation of future mitigation measures so they would meet the Standard. In practice, it has been used for a variety of additional purposes. For instance, a number of developers have used the Standard early in their project planning as a framework for risk assessment, to check whether it will be possible to design the project to achieve NNL/NG, and to identify any potential challenges so they can be addressed very early in the project cycle. In addition, teams working on the development of policies or guidelines related to the mitigation hierarchy, biodiversity offsets and NNL/NG, have referred to the Standard and its accompanying guidelines to inform their work, and it has been used by a variety of groups to learn and raise awareness about mitigation. The BBOP Standard complements other standards on carbon, and water, and guidelines on alleviation of poverty and helps companies show they meet safeguards established by the International Finance Corporation.

4 Other Guidelines and Materials Produced by BBOP

The BBOP Standard provides measurable indicators and offers more clarity for those gauging performance than brief aspirational principles, but it does not tell the whole story. What methods can you use to determine whether the impacts anticipated by a particular project are capable of being mitigated to the extent of achieving NNL/NG? What tools and metrics are most appropriate for measuring loss and gain of biodiversity, and are there worked examples to see? What do some of the technical terms such as 'additionality' mean? To answer these questions BBOP has published a number of other materials, including:

- **Guidance Notes to the Standard on Biodiversity Offsets** notes to help assess whether mitigation measures have been designed and subsequently implemented in conformance with the BBOP Principles, Criteria and Indicators (BBOP 2012b). This gives an interpretation of each Indicator; key questions for assessment; factors to consider in assessing conformance (conformance requirements and situations that are likely to represent causes of non-conformance); as well as related activities from other Indicators.
- **Glossary** a glossary of the terms found in the Standard and also common in methodologies and guidelines related to biodiversity offset design and implementation.
- **Handbooks** on the design and implementation of mitigation measures, including offsets: The Biodiversity Offset Design Handbook provides information, methodologies and tools from which planners can select the approaches best suited to their individual circumstances, including a generic step-by-step process from the initial conception of a development project to the selection and use of mitigation measures, metrics, exchange rules, and suitable offset sites and activities. Appendices summarise various approaches being used in different parts of the world. The Cost-Benefit Handbook sets out ways of integrating people's use and cultural values into biodiversity offset design and implementation, and ensuring that the project and mitigation measures such as biodiversity offsets leave people living in and around the project and potential offset sites no worse off (and preferably better off) as a result. The Biodiversity Offset Implementation

Handbook offers guidance on the implementation of a successful biodiversity offset, such as ensuring that an effective institutional and management structure is in place, that financial flows are sufficient and secured, and that systems are in place to ensure that the offset objectives are achieved.
- **Resource Papers** which clarify a number of issues related to mitigation, offsets and no net loss. These include: No Net Loss of Biodiversity and Loss-Gain Calculations, Limits to What Can Be Offset, Biodiversity Offsets and Stakeholder Participation, and Biodiversity Offsets and Impact Assessment.
- **Case studies** are available from a number of pilot projects from BBOP and other experiences with the mitigation hierarchy, biodiversity offsets and compensation, from 2009 to 2014. Several relate to individual, specific projects. One is a comparative analysis of many different projects, and another is a fictitious worked example drawn from a real example and anonymised.

All these documents are in the public domain, and available from http://bbop.forest-trends.org/pages/guidelines.

5 Recent Developments and Challenges for the Future

Since 2012, there has been a growing commitment by governments, intergovernmental bodies, banks, export credit agencies, individual companies and non-governmental organisations to 'No Net Loss' and a 'Net Positive Impact' on biodiversity. In 2014, 39 countries had existing laws or policies on and 22 were developing them.[2] This number has risen and, depending on the breadth or precision of scope of policy considered, is likely now to lie between 74 and 100 countries.[3] Some 37 companies have also made voluntary commitments related to No Net Loss, Net Gain or related concepts.[4] In addition, the CEOs of 50 companies mainly in the manufacturing and retail sectors who comprise the Board of The Consumer Goods Forum, have pledged to mobilise resources within their businesses to help achieve zero net deforestation by 2020.[5] Since 1 January 2012, the International Finance Corporation (IFC) and 90 members of the Equator Principles Association require clients with impacts on natural and critical habitat to demonstrate no net loss and a net gain of biodiversity, respectively.[6] The World Bank has updated its own safeguard policies related to no net loss[7] and IUCN—The World Conservation

[2]ten Kate and Crowe (2014).

[3]Maron et al. (2016); pers. comm. Eugenie Regan, TBC, January 2017.

[4]ICMM & IUCN (2012). Independent report on biodiversity offsets. Prepared by The Biodiversity Consultancy. Available at: www.icmm.com/biodiversity-offsets

[5]http://www.theconsumergoodsforum.com/strategic-focus/sustainability/board-resolution-on-deforestation

[6]IFC (2012a, b) and http://www.equator-principles.com/

[7]http://www.worldbank.org/en/programs/environmental-and-social-policies-for-projects/brief/the-environmental-and-social-framework-esf

Union has developed a policy on biodiversity offsets[8]. Many of these developments incorporate or draw on the BBOP Principles, and many more professionals are working in this field. In June 2014, BBOP, Defra and ZSL hosted a summit on No Net Loss attended by 300 participants working on these issues from over 30 countries: a gathering that would have been inconceivable 10 years ago.

It is encouraging to see growing attention being paid to the more rigorous application of the mitigation hierarchy. However, enormous changes are needed to improve the quality of mitigation measures, biodiversity offsets, associated laws, policies, guidelines and corporate practices. Furthermore, the lessons learned from experience to date—successes and failures alike—need to be applied at a much greater scale if they are to make a significant contribution to the conservation of biodiversity, to improved land-use planning and sustainable development and to the management of corporate risk and opportunity.

International experience suggests that planning for NNL/NG succeeds when[9]:

- Measures are in place to improve the application of the mitigation hierarchy, and not simply to plan offsets (which should be the last step);
- Clear, consistent policy and guidance is available, for certainty and to avoid delays;
- There are clear roles for national, state and local government and good coordination between government departments (since ministries responsible for planning, land, energy, mining, housing, transport, communities and justice all have a key bearing on long-term land use);
- Performance monitoring and enforcement is ensured through good governance and adequate budgetary provision; clear principles and standards are in place;
- Legal and financial instruments needed to secure long-term implementation are available;
- Proportionate approaches are planned, with more streamlined procedures and simpler baseline studies and metrics for less significant impacts on biodiversity, and full assessments and metrics for more significant impacts;
- There is a realistic roadmap to develop the NNL/NG system and improve it over a few years;
- Preparation for implementation (including supply) takes place during the policy development phase; good baseline data, mapping and landscape level planning are available;
- Methods that don't deliver NNL/NG (e.g. poor metrics) are avoided;
- Several options for implementation are possible, provided the same standards are met;
- Perverse incentives are removed; and
- Assistance is offered to parties such as developers and offset providers who need to find each other.

---

[8]https://portals.iucn.org/library/sites/library/files/resrecfiles/WCC_2016_RES_059_EN.pdf

[9]See ten Kate and Crowe (2014).

Lessons such as these, methodologies, tools and approaches already exist to help policy-makers, developers, conservation scientists and communities plan for NNL/NG and improve rapidly on past performance. That said, more research, experience, analysis and monitoring of outcomes would undoubtedly improve the effectiveness of efforts to plan for NNL. Some of the principal remaining challenges and thus priorities for future work were highlighted recently by IUCN's technical study group on offsets, and are summarised in Box 2.

Box 2 Issues on Which Further Guidance Is Particularly Necessary According to the IUCN Technical Working Group on Offsets (IUCN 2014)

- **Mitigation hierarchy**: Whether there should be information on the manner in which each of the steps within the mitigation hierarchy should be applied and, if so, what that information should be; and whether and how to apply a risk-based approach to the mitigation hierarchy.
- **Dynamic offsets**: How to design offsets within dynamic landscapes that are likely to change during offset duration (e.g. owing to change in other threatening processes, such as population growth or climate change).
- **Site selection and landscape-level planning**: Where to place offsets in relation to impacts, in varying contexts, including when and how to use composite offsets (in more than one location), to address all the biodiversity components impacted by an individual project, or aggregated offsets to cluster together offsets for a number of different projects.
- **Goal**: The appropriate level of ambition for offset policies (e.g. No Net Loss vs. Net Gain).
- **Terminology**: Consistency of use and interpretation of terms such as No Net Loss and Net Gain;
- **Values**: Resolving any conflicts between societal values and 'intrinsic values';
- **Exchange rules**: Establishing exchange rules in order to support conservation priorities, while also ensuring that the offset system runs smoothly;
- **Additionality**: Whether, and if so how, it is possible to demonstrate additionality; How to determine the additionality of activities within existing protected areas, and averting risks in jurisdictions where government policy or investment should already prevent such risks.
- **Baselines**: Best practice in determining the baseline risk of loss for averted risk offsets and in quantifying security gains.
- **Leakage**: Tackling leakage in offset design and implementation.
- **Timing**: When offset gains should be made; The appropriate duration of offsets and how to demonstrate secure long-term offset outcomes when the law does not cater for long-term security of land-use.

(continued)

Box 2 (continued)

- **Standards** needed for implementation (e.g. development and delivery of conservation credits).
- **Roadmaps**: How governments can develop roadmaps for establishing offset systems and market-based approaches to offset implementation.
- **Effective monitoring, evaluation and enforcement** systems.

6 What Future for the BBOP Principles?

The authors hope that the BBOP's Principles will continue to be used and reflected in laws, policies, standards and corporate commitments worldwide. For their own part, the BBOP members intend to:

Update and expand our set of principles, standards, guidelines, tools and case studies in the light of experience from those using them around the world and also to reflect international developments in law, policy, practice and science;

Encourage the sharing of ideas and experience on planning, impact assessment, mitigation and biodiversity offsets and the development of best practice through the growing Community of Practice. This now comprises over 2000 individuals worldwide, and is facilitated by the BBOP Secretariat.

BBOP's current work programme continues at the level of individual sites and projects, but members are also working together to scale up dramatically. We are focussing on high standards and practicality of approaches in planning 'Net Gain' of biodiversity at the jurisdictional level, in government policy at the local, state, national and regional levels. This is particularly through a 'roadmap' to help governments work towards 'Net Gain' meeting typical milestones over a period of years. Similarly, BBOP is developing a roadmap to help companies work towards Net Gain. Members are also developing tools and approaches to integrate Net Gain of biodiversity with other approaches, such as Natural Capital Accounting.

BBOP remains very open to new members, to ideas and suggestions. Anyone with an interest in this field is welcome to take part in webinars and conferences and raise and discuss issues through the Linked In Group.

References

Business and Biodiversity Offsets Programme (BBOP) (2012a) Standard on biodiversity offsets. http://bbop.forest-trends.org/guidelines/Standard.pdf

Business and Biodiversity Offsets Programme (BBOP) (2012b) Guidance notes to the standard on biodiversity offsets. http://bbop.forest-trends.org/guidelines/Standard_Guidance_Notes.pdf

ICMM & IUCN (2012) Independent report on biodiversity offsets. Prepared by The Biodiversity Consultancy. www.icmm.com/biodiversity-offsets

International Finance Corporation (2012a) 1 Performance standard 6: biodiversity conservation and sustainable management of living natural resources. http://www.ifc.org/wps/wcm/connect/c8f524004a73daeca09afdf998895a12/IFC_Performance_Standards.pdf?MOD=AJPERES. 1 Jan 2012

International Finance Corporation (2012b) Guidance notes to performance standards on environmental and social sustainability – 2012 edition. http://www.ifc.org/wps/wcm/connect/topics_ext_content/ifc_external_corporate_site/ifc+sustainability/publications/publications_policy_gn-2012

IUCN (2014) Biodiversity offsets technical study paper. IUCN, Gland

Maron M, Christopher DI, Kujala H, Bull JW, Maseyk FJF, Bekessy S, Gordon A, Watson JEM, Lentini PE, Gibbons P, Possingham HP, Hobbs RJ, Keith DA, Wintle BA, Evans MC (2016) Taming a wicked problem: resolving controversies in biodiversity offsetting. Bioscience 66 (6):489–498. https://doi.org/10.1093/biosci/biw038

ten Kate K, Crowe MLA (2014) Biodiversity offsets: policy options for governments. An input paper for the IUCN Technical Study Group on biodiversity offsets. IUCN, Gland

Austria

Martina Artmann

Austria is a landlocked country of 83,855 km^2 situated in Central Europe. Its neighbours are Germany and the Czech Republic to the north, Slovakia and Hungary to the east, Italy and Slovenia to the south and Switzerland and Liechtenstein to the west. In 2015 the population was 8,579,747 (Statistik Austria 2015a), a figure expected to increase to 9,194,135 inhabitants by 2030 (Statistik Austria 2015b). Austria is a federal republic made up of nine autonomous federal states: Burgenland, Carinthia, Lower Austria, Upper Austria, Salzburg, Styria, Tyrol, Vorarlberg as well as the city-state and Austrian capital, Vienna.

Twenty-seven percent of Austria's territory is protected by environmental legislation, of which 16% is composed of strictly-protected NATURA 2000 sites, national parks or nature reserve areas (Umweltbundesamt 2015a). However, the current biodiversity monitoring report issued by the Austrian Federal Ministry of Agriculture, Forestry, Environment and Water Management (BMLFUW) warns of the threat to the country's species and habitat diversity. Some reasons for the current pressures on biodiversity are a decrease in upland farming, an increase in the fragmentation of agricultural land as well as the negative impacts on ecosystems by ongoing land take and land use for human settlements and transport infrastructure (Fiala 2014). To deal with this situation, the "Biodiversity Strategy Austria 2020+" was developed to support the EU Biodiversity Strategy. Objectives of the strategy are to conserve local biodiversity and to reduce the loss and deterioration of species, habitats and genetic diversity. In the strategy there are no explicit links to the EU-wide No Net Loss initiative (NNL) mentioned. Only implicitly one link to strategies dealing with impact mitigation is considered by recommending a functional relationship between compensation areas supporting a habitat network (BMLFUW 2014).

M. Artmann (✉)
Leibniz Institute of Ecological Urban and Regional Development, Dresden, Germany
e-mail: m.artmann@ioer.de

W. Wende et al. (eds.), *Biodiversity Offsets*,
https://doi.org/10.1007/978-3-319-72581-9_4

This chapter presents the regulations for impact mitigation in Austria (see also Süssenbacher and Jungmeier 2017). In the following, the term "impact" is used as a general term to describe possible effects of interventions and interference in nature and landscape.

1 Theory in Austria

1.1 Legal Background in Austria

Nature and ecosystem conservation in Austria is regulated by several material laws, e.g. forestry law, water law and nature conservation law. In addition to the division of concerns relevant for biodiversity and ecosystem conservation between the material laws, the legislative competency also varies, i.e. while water law and forest law is under federal jurisdiction, game law, laws on fisheries and nature conservation law are the responsibility of the state governments. The impact mitigation regulation plays, at least implicitly, a role in the different material laws, but can be explicitly regulated by the nature conservation laws of the Austrian states.

The environmental impact assessment (EIA) requirement under the Austrian Environmental Impact Assessment Law 2000 (UVP-G 2000) is the only legal measure that applies relevant material laws in one concentrated procedure. Besides the nature conservation laws, the UVP-G 2000 explicitly provides for impact mitigation regulation. Thus, in Austria there are two main legal bases for impact mitigation regulation: (a) nature conservation legislation and (b) UVP-G 2000, both of which are presented below.

1.1.1 Regulations by nature conservation legislation in Austria

Under Austria's federal system, the legislation and implementation of nature protection falls under the jurisdiction of the federal states. This means that rather than some universally applied national law on nature protection, there exist nine different nature protection laws set by the states (see Table 1). Consequently, there are also nine different regulations of the impact mitigation procedure required under nature conservation law. These differ substantially in detail and quality between the various states. This is in contrast to Germany where the impact mitigation regulation (*Eingriffsregelung*) is firmly anchored in a framework law at the federal level by the German Federal Nature Conservation Act before being further refined according to the law of each federal state (see chapter "Germany").

Impacts on the ecosystem and landscape are regulated at project level by the federal nature protection laws via permit requirements and duty of disclosures. The nature protection laws dictate which impacts need permit requirements and which preconditions related to targets of nature conservation and landscape development need to be kept for permission to be granted. Although similarities can be found

Table 1 List of federal nature conservation laws in Austria

Federal state	Nature conservation law	Abbreviation
Burgenland	Burgenland Nature Conservation and Countryside Protection Law 1990	Bgld NSchG
Carinthia	Carinthian Nature Conservation Act 2002	Krnt NSchG
Lower Austria	Law of Lower Austria on Nature Protection 2000	NÖ NSchG
Upper Austria	Upper Austria Nature and Countryside Protection Act 2001	OÖ NSchG
Salzburg	Law of Salzburg on Nature Conservation 1999	Sbg NSchG
Styria	Styrian Nature Conservation Act 1976	Stmk NschG
Tyrol	Tyrolean Nature Conservation Law 2005	Tir NSchG
Vorarlberg	Law on Nature Conservation and Landscape Management 1997	Vlbg NSchG
Vienna	Vienna Nature Conservation Act 1998	W NSchG

between the specific impacts needing permissions and the related approval preconditions, the acts differ between the states regarding the formulation, detail and structure of the separate clauses. The clauses vary in regard to the regulation of *types of impacts* (e.g. construction of buildings with a specific height or construction of roads with a specific width) as well as the protection of the *area of the impact* (e.g. protected areas such as landscape reserves, water protection, riparian forests) (Bußjäger 2013).

If impacts conflict with nature conservation and landscape targets, permissions are generally not granted (principle of avoidance). For instance, permissions have to be refused if there are negative impacts on the natural scenery, character of the landscape or structure of the natural balance (§ 6 Bgld NSchG) or on the balance, shape and recreational function of the landscape (§ 18 para. 3 W NSchG). However, if public interests prevail over nature conservation interests, the project may be approved. As stated in § 6 para. 5 Bgld NSchG, public interests include, among other things, national defense, environmental protection, the economy and tourism, land reform and agriculture. If public interests are more valuable than those of nature conservation, approvals for development can be issued by the competent authority. If the principle of avoidance cannot be kept, approvals are only granted if impacts are minimized as far as possible according to all federal nature conservation acts (§ 6 para. 6 Bgld NSchG; § 9 para. 8 Krnt NSchG; § 7 para. 2 NÖ NSchG; § 14 para. 2 OÖ NSchG; § 2 para. 2 lit. b Sbg NSchG; § 5 para. 7 Stmk NschG related to nature protection areas; § 29 para. 5 Tir NSchG; § 37 para. 1 Vlbg NSchG; § 11 para. 4 W NSchG). Thus, the Austrian nature conservation acts apply a basic mitigation hierarchy. The environmental impact assessment includes a cascade model: (1) avoidance, (2) minimization and (3) compensation (Darbi et al. 2010).

If further impacts are necessary due to some overriding public interest, requirements can be set by prescribing conditions, requirements or time limits for compensating the impacts on the ecosystem and landscape as the last step of the cascade model. These requirements form, at least implicitly, the basis for prescribing compensation or offset measures. However, some federal states also explicitly provide regulations on impact mitigation measures. Figure 1 presents an overview of the

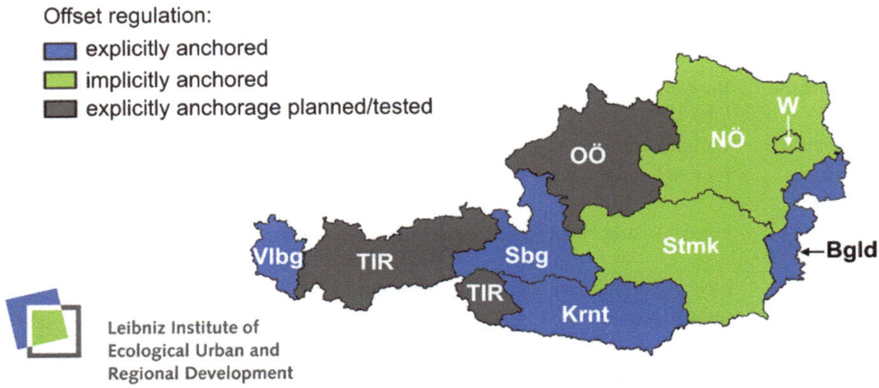

Fig. 1 Implementation of offset regulations in the Austrian federal nature conservation acts (own presentation, state boundaries based on data.gv.at 2015). *Bgld* Burgenland, *Krnt* Carinthia, *NÖ* lower Austria, *OÖ* upper Austria, *Sbg* Salzburg, *Stmk* Styria, *Tir* Tyrol, *Vlbg* Vorarlberg, *W* Vienna

state and extent to which impact mitigation regulations are embedded in the nature conservation law of each Austrian federal state.

Four Austrian states have explicitly implemented legal clauses for impact mitigation regulation, i.e. Burgenland, Carinthia, Salzburg and Vorarlberg. Table 2 gives the names of the related clauses in which the impact mitigation regulation is placed and provides an overview of the formulation of the different clauses in the states. In all four states, the impact mitigation regulation includes the compensation of habitats for flora and fauna. Only Burgenland and Salzburg refer to offsets mitigating impacts on the landscape such as its scenery or recreational function. The focus of habitat compensation is on the creation of habitat substitutes whereas all states allow the possibility of monetary compensation when the creation of natural habitat substitutes is impossible or unreasonable (see Table 2).

Regarding natural compensation, Salzburg has the most specific requirements in Austria. These are specified in two clauses regarding impacts of public (§ 3a Sbg NSchG) and private interests (§ 51 Sbg NSchG). Permissions for impacts in the course of private overriding interests are possible if compensatory measures (German: *Ausgleichsmaßnahmen*) are undertaken that lead to significant improvements in the landscape or the natural balance (§ 51 Sbg NSchG). For private projects, it is also possible to credit compensation measures already implemented under the condition that—as with newly created compensation measures—they significantly improve the landscape or the natural balance. Only existing measures implemented less than 3 years before the application can be credited, or in exceptional cases, measures introduced up to 6 years previously (§ 51 para. 2a Sbg NSchG). § 3a Sbg NSchG is applied when public interests override; however, resulting impacts entail substitute remediation, especially in the form of substitute habitats. As far as possible these should be spatially related to the area of impact (§ 3a para. 4 Sbg NSchG).

Table 2 Explicit anchorage of impact mitigation in Austrian nature conservation acts that goes beyond the EU Habitats Directive requirements

State	Explicit legal anchorage of impact mitigation	Legal guidelines for impact mitigation
Burgenland	Compensation of ecological impacts (§ 10 Bgld NSchG)	1) Provisioning of a suitable substitute habitat if habitats of rare, endangered or protected animals and plants are considerably affected or destroyed by the impact (§ 10 para. 1a Bgld NSchG). 2) Furnishing of compensation when impacts significantly and sustainably harm scenic nature, the landscape character, the beauty or the recreational value of a landscape part (§ 10 para. 1b Bgld NSchG). 3) If a stipulation for habitat substitute is impossible or unreasonable, a cash amount equal to the cost of creating a suitable habitat replacement has to be paid (§ 10 para. 2 Bgld NSchG).
Carinthia	Habitat substitute (§ 12 Krnt NSchG)	1) If habitats of rare, endangered or protected animals and plants are considerably affected or destroyed by the approved impact, a suitable habitat substitute needs to be created (§ 12 para. 1 Krnt NSchG). 2) If such a stipulation is impossible or unreasonable, a cash amount equal to the cost of creating a suitable habitat replacement has to be paid (§ 12 para. 2 Krnt NSchG).
Salzburg	a) Substitute remediation: if public interests override nature conservation interests, expected impacts need to be compensated by substitute remediation (§ 3a Sbg NSchG) b) Compensatory measures: if private interests override nature conservation interests, expected impacts need to be compensated by compensatory measures (§ 51 Sbg NSchG)	ad a) Regarding impacts in special habitats and communities of animals or plants, the priority of compensation needs to be the creation of habitat substitutes a.1) preferably in the immediate vicinity of the affected area. a.2) If the creation of habitat substitutes is impossible, a cash amount equal to the cost of creating a suitable habitat substitute has to be paid. ad b) Instead of prohibiting an impact, the authority can prescribe or credit compensation measures (§ 51 para. 1 Sbg NSchG) if all conditions are fulfilled (§ 51 para. 3 Sbg NSchG): b.1) Compensatory measures need to lead to a significant improvement of the landscape or natural balance. b.2) This improvement outweighs the adverse effects in the affected or neighboring areas.

(continued)

Table 2 (continued)

State	Explicit legal anchorage of impact mitigation	Legal guidelines for impact mitigation
Vorarlberg	Time limits, conditions, requirements (§ 37 Vlbg NSchG)	1) Conditions and requirements may include stipulations for ecological compensation measures such as habitat substitutes (§ 37 para. 3 Vlbg NSchG). 2) If the creation of habitat replacement is impossible, a cash amount equal to the cost of creating a suitable habitat substitute by the state has to be paid (§ 37 para. 3 Vlbg NSchG).

In general, Salzburg has played a pioneering role in Austria in the application of the impact mitigation regulation, which was first introduced in 1993. This state provides objective guidelines on how to assess impacts in nature and landscape and how to determine the need for compensation. Thus, Salzburg serves as a role model for other Austrian states in monitoring the quality of compensation measures (Jungmeier et al. 2010; Knollconsult Umweltplanung et al. 2016) (see Sect. 2).

Although not all Austrian federal states have explicit legal regulations on impact mitigation, current discussions by practitioners highlight its growing importance, as shown by the sixth expert panel of the Austrian Federal Forestry Office in 2013 on the topic “Compensation areas—challenges for successful implementation” (for overall outcomes, see Österreichische Bundesforste 2014). Moreover, there are increasing efforts by other states such as Upper Austria and Tyrol to develop and test compensator mitigation regulations (see Fig. 1).

In 2014, the Upper Austria Nature and Countryside Protection Act was revised to allow for the approval of compensatory measures (a) for unavoidable impacts on areas which are valuable from a nature conservation perspective and (b) for impacts of structural and ecological habitat functions of protected plants and animals. To this end, a separate regulation was developed to assess impacts in nature and landscape for the determination of appropriate compensation measures. This regulation came into force in April 2015. The first year was intended to be a testing period for the regulation and applicants alike, with results being used to introduce refinements (Schiffner and Matzinger 2015). Although it has already established explicit impact mitigations, Upper Austria basically rejects monetary compensation (Land of Upper Austria 2012). In Tyrol, where currently impact mitigations are indirectly regulated by permit requirements, the development and implementation of an explicit legal clause is planned (Sladek 2013; s.n. 2010).

In addition to the explicit or implicit regulation of impact mitigation in nature conservation laws in several Austrian states, each state codifies compensation measures for European protection areas in its nature conservation law (§ 22d para. 3 Bgld NSchG; § 24b para. 2 Krnt NSchG; § 10 para. 7 NÖ NSchG; § 24 para. 6 OÖ NSchG; § 3a para. 5 Sbg NSchG; § 13b para. 5 Stmk NSchG; § 14 para. 6 Tir NSchG; § 26 para. 4 Vlbg NSchG; § 22 para. 7 W NSchG). Thus, according to

§ 6 para. 4 of the EU-Fauna-Flora-Habitat Directive, compensation measures are required if, despite a negative impact assessment, impacts on nature are necessary due to overriding public interests (including those of a social or economic nature) and a lack of alternatives. The compensatory measures must ensure that the overall coherence of Natura 2000 is preserved.

1.1.2 Regulations by the Environmental Impact Assessment Law (UVP-G 2000) in Austria

Alongside legal regulations for impact mitigation measures under the federal nature conservation laws, the Austrian Environmental Impact Assessment Law 2000 (UVP-G 2000), first enacted in 1993 and revised in 2000, is an important tool to regulate impacts on nature (Peters et al. 2002; Roskosova 2014). Compulsory UVP projects are approved via an expedited permit procedure. For the realization of projects, the federal state applies all relevant laws. Thus, impact mitigation regulations codified in the nature conservation laws feed into the UVP-G. Projects requiring an EIA are regulated by the Austrian UVP-G 2000 in Annex 1 as well as in Annex 3, where regulations for federal roads and high-speed rail segments describe no fewer than 89 project types (see Fig. 2) including the various thresholds for new projects, projects involving a change of use and projects in protected areas.

For the derivation of compensation measures by the UVP-G 2000, the environmental impact statement [German: *Umweltverträglichkeitserklärung* (UVE)] is of particular importance (§ 6 UVP-G 2000). The aim of the UVE is to specify all environmentally relevant project data including the actual state of the affected environment, types and degree of impacts on the environment as well as the most viable implementation alternatives. According to § 6 para. 5 UVP-G 2000, the project applicant must also provide details of measures to avoid, minimize and compensate for any adverse impacts on the environment. Thus, there is an immediate relevance for compensation measures under the basic mitigation hierarchy.

Based on the UVE, comments and reviews of the potential project impacts on the environment are provided by experts from various special fields appointed by the responsible authority. The evaluation feeds into an extensive environmental impact audit [German: *Umweltverträglichkeitsgutachten* (UVG)]. In the UVG, suggestions on avoiding, minimizing and compensating for environmental impacts are evaluated and, if necessary, augmented. After an oral hearing, the responsible authority makes its decision. According to § 17 para. 4 and 5 UVP-G 2000, a favorable decision is only granted upon proof of sufficient planned compensation measures that would achieve a high level of environmental protection. Thus, the UVP-G 2000 strengthens the nature conservation legislation, in particular in states with weak regulations for compensation measures. However, the UVP-G 2000 also limits the legitimacy of impacts when a high level of environmental protections cannot be secured by compensation measures (Klaffl et al. 2006). In the case of a positive decision, the project applicant is also informed of the compensation measures which have to be implemented as a part of the project realization. In the course of the UVP-G 2000,

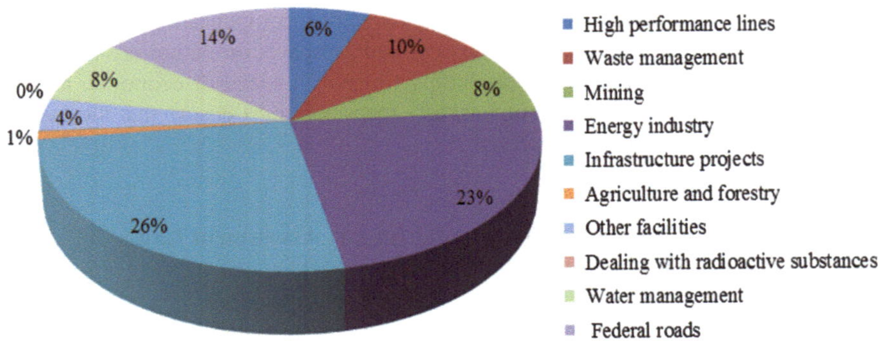

Fig. 2 Compulsory UVP project types according to UVP-G 2000 (as of May 2014) (own diagram based on data provided by the Umweltbundesamt 2015b)

public access to information and/or participation is ensured in several stages of the process. Between 2009 and 2014, 193 approval procedures under the UVP-G 2000 were initiated in Austria (Umweltbundesamt 2015c), of which 29 applications were accepted and five rejected (Umweltbundesamt 2015d).

1.2 Methodological Background in Austria

1.2.1 Metrics in Austria

As stated in Sect. 1, the regulation of impact mitigation compensation in Austria can for the most part be legally applied under the umbrella of the federal nature conservation laws and the UVP-G 2000. Different guidelines are provided for these types of regulations to evaluate impacts and compensation. The guidelines are presented in this section.

Metrics According to Austrian Nature Conservation Laws Using the Example of Salzburg

Although four Austrian states have explicitly anchored the application of compensation measures to their federal nature conservation laws (see Fig. 1), only Salzburg provides detailed, objective guidelines on how to evaluate and determine project impacts on the environment and landscape and their related compensation requirements. Thus, the focus of this section is on the "Guidelines on Preparing Reports for Nature Conservation with regard to the Assessment of Replacement and Compensatory Measures according to the Salzburg Nature Conservation Act" by Loos (2006). The guidelines were developed by the author together with spatial and landscape planners, the environmental authority of Salzburg and a range of experts and commissioners for nature conservation of the state government office to improve the traceability in the assessment of impacts and compensation measures (Loos 2006).

As stated in Table 2, the legal regulation of impact mitigation in Salzburg prescribes compensatory measures for projects with overriding *private* interests (§ 51 Sbg NSchG) and substitute remediation for impacts from projects with overriding *public* interests (§ 3a Sbg NSchG). The guidelines provided can be applied to both kinds of regulations. Although the Sbg NSchG makes some distinctions between compensatory measures and substitute remediation, the guidelines state that the term "compensation" also encompasses "substitute remediation" (Loos 2006). Thus, in the following text "compensation" is also used to mean "substitution" and "offset".

Basically, according to the guidelines, impacts and compensation are calculated separately whereas the impact and compensation related to the ecosystem and landscape (incl. its recreational function) are considered individually. Moreover, the guidelines reflect on the duration of the effects of the impact and the compensation measure and recommend that compensation measures should stand in a close spatial relationship with the impact (e.g. in the same municipality, natural environment or landscape zone). The evaluation is realized by value scores (VS).

The basis for calculating the VS for impacts and compensation on the ecosystem is the detection of the sizes of the affected areas before and after the impact and compensation, which are multiplied by a VS related to the ecosystem (VS_E). The VS_E is determined by assigning the areas to biotope and land use types. These are ranked by their importance in terms of nature conservation, ranging from 0 (no importance) to 6 (most important). The following criteria are considered when classifying the areas according to their significance for nature conservation: rarity, naturalness, vulnerability, development time and functionality in the ecosystem. New biotopes and land use types to be developed such as listed in Table 3 are valued less than already existing ones. The guidelines provide then a list of the most common biotopes and land use types and their classification in the VS_E (see Table 3).

The product of area size and its related VS_E before the impact/compensation is subtracted from the product of area size and its related VS_E after the impact/ compensation. If an area affected by impact and compensation consists of different VS_E, the scores of all areas are summed up. Adjustments may be made afterwards depending on the effect duration of the impact and/or compensation (see Table 4).

In the course of the evaluation of the compensation, another correction factor is included to consider the timing of the compensation implementation. The effect duration of the impact covers the period from the legal start of the impact until the end of the right conferred. The effect duration of the compensation usually starts at the point of completion of the compensation measure and ends with the completion of the obligation imposed according to the notification. If the start of the implementation of the compensation is later than 1 year after the implementation of the impact, an additional correction factor must be applied which adjusts the delayed timing of the compensation implementation (see Table 4). Figure 3 provides a schematic representation of the calculation of impacts and compensation on the ecosystem.

Similar to the calculations referring to the ecosystem, the basis for calculating the VS for impacts and compensation on the landscape (L) is the detection of the sizes of the affected area before and after the impact and compensation and their multiplication by a VS related to the landscape (VS_L). With the VS_L, the scenery and the

Table 3 Examples of value scores (VS) for common biotopes and land use types in Salzburg and their exemplified classifications VS_E for the ecosystem (E) [own table based on guidelines by Loos (2006)]

VS Type	0	1	2	3	4	6
Forests	x	Non-native forests, productive orchards, Christmas tree plantations	Avenues in cultural landscapes, single trees and patches of woodland	Mixed orchards, for local landscape important single trees	Near-natural and structurally rich forests and hedges	From a nature conservation perspective, most valuable primeval and natural forests
Arable land	x	Intensively cultivated fields	Extensively cultivated fields	x	x	x
Standing waters	Basin without natural features	Artificial, poorly structured pristine and bathing lakes	Near-natural and well-structured pristine and bathing lakes	Near-natural still waters with natural trophic status	Natural still waters	Mostly natural still waters of highest level of protection
Biotopes and land use types in settlement area	Developed land and sealed surfaces	Gardens, cemeteries, playgrounds, park and sport lawns	Parks with old trees	x	x	x

Table 4 Correction factors to take account of time dimensions of impacts and compensation [own table based on guidelines by Loos (2006)]

Correction factor	Effect duration of the impact and compensation
1.0	Over 20 years
0.8	16–20 years
0.6	11–15 years
0.4	6–10 years
0.2	5 years or less
Correction factor	Delayed timing of the compensation implementation
1.0	Simultaneously or max. 1 year after the intervention
0.9	Max. 3 years after the intervention
0.8	Max. 5 years after the intervention

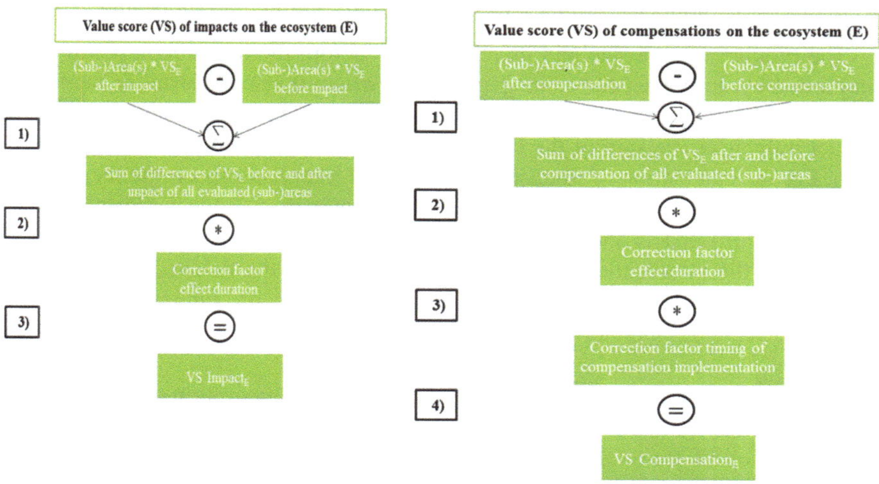

Fig. 3 Calculation of VS of impacts and compensation on the ecosystem (E) [own diagram based on Loos (2006)]

character of a landscape is evaluated on a scale of 0–6, similar to that of the VS_E (0 = no importance, 6 = most important). The following examples provide a rough breakdown of such value scores:

- 0: landscapes of metropolitan character influenced by spacious industrial and commercial areas or infrastructure
- 1: widely sprawled landscapes; landscapes are structurally poor and intensively used for forest, agriculture and settlements
- 2: settlements and cultural landscapes with an average supply of cultural landscape elements

Table 5 Supplement factor to evaluate impacts of interventions on the recreational landscape function [own table based on guidelines by Loos (2006)]

	Impact of the intervention on the recreational value of the landscape				
Recreational value of the landscape	Negligible	Small	Medium	High	Very high
Negligible	1.0	1.0	1.0	1.0	1.0
Small	1.0	1.1	1.2	1.3	1.4
Medium	1.0	1.2	1.4	1.6	1.8
High	1.0	1.3	1.6	1.8	1.9
Very high	1.0	1.4	1.8	1.9	2.0

- 3: traditionally-used cultural landscapes with above-average, high-quality supply of cultural landscape elements
- 4: high-quality cultural landscapes of a special natural beauty and/or character
- 6: unique natural and cultural landscapes of outstanding natural beauty

The product of area size and its related VS_L before the impact/compensation is multiplied by an impact factor in steps of 0.2, where 0 stands for no or negligible impacts and 1 for extraordinary impacts in the related landscape zone. The type of impact influences the sign of the factor (impact = negative; compensation = positive). For the calculation of impacts, the impact of interventions on the recreational landscape function is specified by a supplement factor (see Table 5). Similar to the calculations of impacts and compensation on the ecosystem, a factor for the effect duration of the intervention and compensation and for the delayed timing of the compensation implementation is considered (see Table 4). Figure 4 provides a schematic representation of the calculation of impacts and compensation on the landscape.

Based on the separate calculations for impacts and compensation measures on the ecosystem (see Fig. 3) and the landscape (see Fig. 4), Fig. 5 provides an overview of the calculation of the total value scores. Depending on the balancing of private or public interests, the compensation of the impact is achieved when:

- According to § 3a Sbg NSchG, the ratio between substitute remediation and compensation is 1:1 (consideration of public interests) or
- According to § 51 Sbg NSchG, the ratio between impact and compensation is 1:1.3 (consideration of private interests).

Metrics According to UVP-G 2000

For project applicants, the UVE is an elementary component to evaluate impacts on nature and the environment by the planned project, its investigated alternatives and to determine offset measures based on the evaluation results (see Sect. 1.1). Although there are a range of guidelines on the development and process of a UVE, for instance for commercial and industrial areas (Baumgartner and Eberhartinger-Tafill 2006a), skiing areas (Baumgartner and Eberhartinger-Tafill 2006b) and factory farming (Büchele and Eberhartinger-Tafill 2006), they lack

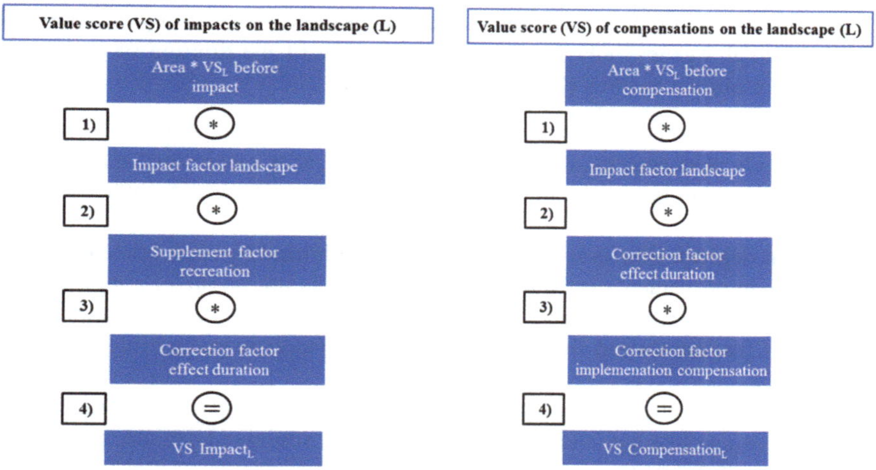

Fig 4 Calculation of value scores (VS) of impacts and compensation on the landscape (L) [own diagram based on Loos (2006)]

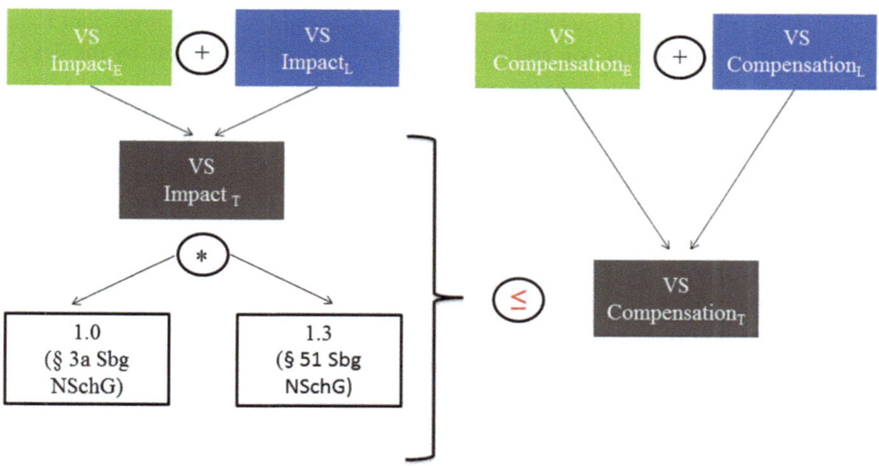

Fig. 5 Calculation of total (T) value scores (VS) of impacts and compensation [own diagram based on Loos (2006)]

legally binding standards for the quality and quantity of compensation measures for the UVP, such as the drafting of reforestation measures (Österreichischer Rechnungshof 2008; Roskosova 2014). In addition, guidelines on how to draft the

UVE and its required descriptions of compensation provided by the Federal Environmental Agency only give a rough overview of the drafting of mitigation and compensation measures (Umweltbundesamt 2012):

- **Habitat destruction** should be compensated by the development of new habitats for the affected species at another location and to a satisfactory quality. If a suitable replacement is not possible, an equivalent compensation offset shall be implemented.
- **Habitat degradation** should be mitigated or minimized by appropriate measures (e.g. by improvements in water ecology). If this is not possible, a replacement equal to that which is impacted through development shall be implemented.
- **Habitat fragmentation** shall be mitigated by corridors or green bridges/wildlife crossings.
- **Changes in species** can be mitigated or promoted by setting a range of measures (e.g. planting of hedges and bushes to create richly structured habitats, creation of near-natural still waters).

Within the guidelines no specific methods and value scores for assessing the effectiveness of the mitigation and compensation measures are included. However, it is stated that a transparent evaluation matrix has to be defined for assessing the impact of the measures (Umweltbundesamt 2012). In the UVP, one important issue is the minimization, mitigation and compensation of negative impacts by the development of traffic infrastructure (see Sect. 1.1). An efficient determination of offsets by the UVP in the road and traffic sector has become even more important due to the expansion in expressways and motorways in most of Austria's states between 1995 and 2012: for instance, 73 km extra roadway in Upper Austria, 138 km in Carinthia and even 291 km in Burgenland (BMLFUW 2013). In particular, the increase in traffic infrastructure and its related sealing of landscapes threatens local habitats and biodiversity (BMLFUW 2009).

To improve the details, quality and transparency in determining compensation measures for the traffic sector, standardized guidelines on mitigation and compensation are currently being drawn up by the Austrian Research Association Roads-Railways-Traffic (FSV) (FSV 2015). These guidelines will be part of the "Guidelines and Rules for the Road Sector (RVS)" provided by the FSV. These are a legally binding set of rules for federal or country roads that, based on the latest technologies, provide standards for regulating environmental impacts (section RVS 04) in the course of the development of projects for the road sector, e.g. guidelines on amphibian protection near roads (RVS 04.03.11), game protection (RVS 04.03.1), protection of birds near roads (RVS 04.03.13), wild mammal conservation near roads (excluding bats) (RVS 04.03.14).

In addition to these sector-based guidelines, metrics on impact mitigation regulations provided by the natural conservation laws can feed into the UVP. Thus, the UVP is a concentrated procedure that integrates relevant material laws.

1.2.2 Methodology of Habitat Banking

With the UVP-G (2000) and some federal natural conservation laws, impact mitigation regulations have already arrived in Austrian practice. However, the concept of habitat banking, mitigation banking or eco-accounts has yet to be implemented. Practitioners warn that in order to ensure the quality of compensation measures during the project development and implementation, land availability needs to be secured (Bergthaler and Sladek 2014). At the national level, the Federal Environmental Agency plans to submit a European LIFE Project focusing on the reduction of soil loss in Austria. One main topic of the project should be the development of an assessment system to evaluate compensation for the loss of soil functions as well as to define appropriate compensation areas through a regional mitigation bank (Regionalmanagement OÖ 2015).

In general, Austrian stakeholders often refer to the German model *Ökokonto* (eco-account) (see chapter "Germany") as a best practice example and potential role model for the country to improve the quality of its compensation scheme (Bergthaler 2013; Plattner 2014; Proksch and Graf 2011). When developing a mitigation bank for Austria, it is important to consider the quantity of available land as well as its quality regarding the suitability to function as compensation land (Bergthaler and Sladek 2014). Particularly for large infrastructure projects, a mitigation bank could be helpful in providing sufficient compensation land, which often needs to be additionally purchased by applicants to gain a project's approval. This often leads to high costs for compensation measures, especially for impacts on valuable and protected natural areas (Österreichischer Rechnungshof 2008), while also undermining the coordination required to achieve regional targets of nature conservation and landscape planning (Proksch and Graf 2011).

Against this background, a regional study was conducted to develop and test a landscape account in the region of Vienna and Lower Austria, based on the concept of the German eco-account (Proksch and Graf 2011). The highly dynamic agglomeration of Vienna-Lower Austria faces high demand for available land, resulting in land use conflicts such as for settlement, transport and commercial areas as well as for green areas for recreation and nature conservation. Thus, the landscape account should guarantee that the regional development of new unavoidable infrastructure projects and related, legally required compensation measures will take place in accordance with regional landscape planning targets. Moreover, the concept of a landscape account should foster an efficient allocation of funds, which are generally scarce for regional landscape planning measures. To ensure an efficient implementation, the focus of the concept is the establishment of a communication and information platform providing information on regional pilot projects on landscape planning and nature conservation, compensation land availability and contacts for project applicants to the relevant stakeholders (Proksch and Graf 2011).

1.2.3 Offset Quality Assurance in Austria

The variations in the nature conservation laws between the Austrian states also applies to their embedded prescription of quality saving of offsets. As prescribed by nature conservation law, quality saving of compensation measures can generally be secured in monetary terms and by competent monitoring. Regarding the latter, specific monitoring of the implementation of projects and consideration of environmental requirements can be achieved through an ecological building supervision (German: *ökologische Bauaufsicht*). The potential consideration of an ecological building supervisor is regulated in Carinthia, Upper Austria, Salzburg, Tyrol and Vienna (§ 47 Krnt NSchG; § 42a OÖ NSchG; § 50 para. 3 Sbg NSchG; § 44 para. 4 Tir NSchG; § 20 W NSchG). The ecological building supervisor has the responsibility to monitor and supervise projects with environmental impacts. For instance, according to § 42a para. 2 OÖ NSchG, the ecological building supervisor has to advise the project applicant in meeting official environmental prescriptions and implementing compensation measures as required. However, according to Austrian practitioners, the basis for monitoring the correct implementation of compensation by authorities is the setting of clear and measureable targets related to offsets (Bergthaler and Sladek 2014). This is still missing in most nature conservation laws in the various states (see also Sect. 3).

Some federal nature conservation laws regulate the implementation of nature conservation funds and/or landscape maintenance funds, such as in Salzburg (§ 60 Sbg NSchG), Styria (§ 29 Stmk NschG), Tyrol (§ 20 Tir NSchG) and Vorarlberg (§ 10 Vlbg NSchG). The funds are *inter alia* supplied through revenues gained in the course of monetary compensation. For instance, the Regional Court of Audit of Vorarlberg states that revenues through compensation measures must be used to create substitute habitats (Landes-Rechnungshof Vorarlberg 2008). In general, the funds target long-term, financial security of nature and landscape conservation measures and are specifically intended to cover any costs related to offset measures (see e.g. § 20 para. 2 lit. b Tir NSchG). Moreover, Austrian nature conservation laws allow for security deposits to be demanded. For instance, Salzburg has explicitly implemented the option to require security deposits. These have to be paid by the project developers in advance to secure the punctual and complete fulfillment of compensation and offset measures (§ 44 Sbg NSchG).

In the case of a positive decision of a UVP, an acceptance test is conducted after project completion as well as a follow-up inspection after 3 to 5 years. However, there is no provision for the long-term monitoring of compensation measures (see also Sect. 3). Nevertheless, an Austrian survey investigating the performance of UVP projects found that because of its compulsory character, the UVP supports a more integrative assessment of environmental impacts than is the case with separate material laws. The surveyed stakeholders mentioned that innovative compensation measures for nature protection and ecology are initiated because of the UVP (Klaffl et al. 2006). The added value of the UVP compared to the regulations of the federal nature conservation laws is its consistent national regulation ensuring a

UVE through § 6 UVP-G 2000. Thus, the UVE requires the presentation of measures to avoid, minimize and compensate adverse impacts on the environment as well as potential alternatives. Another benefit of the UVP is that compensation measures have to be taken into account at the earliest stages of project planning. In contrast, compensation measures under nature conservation laws are usually not already considered during the submission of the project proposal by project applicants but if barriers arise for approvals during the administrative proceedings (Klaffl et al. 2006).

2 Practice in Austria

Based on the theoretical sections of legal regulations in the two nature conservation laws in Salzburg and the Austrian UVP-G 2000, a project is described here that was approved by the state of Salzburg in 2015 according to the UVP-G. The project applied for the expansion of mining activities to extract mineral deposits at a hard rock quarry in the Salzburg Pinzgau region. Currently the quarry is 16 ha in size, surrounded by a mountain landscape and located next to the River Saalach. Plans called for an 18.5 ha expansion into the hillside to access new mineral deposits, thereby increasing the quarry size to 34.5 ha (see Fig. 6). The successive extension of

Fig. 6 Aerial photo of the project area (red: area of extension; yellow: present quarry). Source: photo provided by DI Markus Ramler, orthophoto: Salzburg State Government, project area: plan creation by FRIEDL ZT GmbH

the raw material extractions includes a follow-up recultivation over a period of 102 years to the final recultivation measure.

The following section provides a summary explaining (1) the legal regulations governing the project, and (2) how impacts and compensation on the ecosystem and landscape were evaluated based on the guidelines of Salzburg (Loos 2006, see Sect. 1.2) and how the implementation and quality of the compensation is guaranteed.

2.1 *Legal Regulations for Project Offsets*

According to the UVP-G 2000, the project has to conduct an environmental impact study (UVE, see Sect. 1.1), in which the project applicant is responsible for evaluating impacts of the project on nature conservation, fauna and flora, biotopes, ecosystems and landscape. Since the project has submitted targets such as the extraction of mineral resources and the building of necessary extraction facilities, it needs permission according to the Salzburg nature conservation law (§ 25 para. 1 lit. a Sbg NSchG 1999). Thus, the project can only be approved if it does not significantly affect the ecosystem, the landscape, its character and recreational function and if there are no suitable alternative solutions with less impact on the ecosystem and landscape.

Based on the UVE submitted by the project applicants, auditors concluded from the UVG that the project will have considerable adverse effects. These impacts are directly related to the quarry expansion and its land consumption, the deterioration of habitats, vegetation and the natural floor structure and consequent impacts on flora and fauna, the fragmentation of nature and landscape, and thus the lowering of scenic and recreational value. Nevertheless, according to the nature conservation law, the project could still be approved since its realization is linked to important public interests, because the expansion of the quarry will guarantee the supply of mineral deposits in the long term, which is of regional and supra-regional importance. Therefore, the project could be approved provided that offsets are implemented that benefit the ecosystem and landscape in a relation of 1:1 (§ 3a Sbg NSchG).

In order to obtain the desired project approval, the project applicants suggested several offset measures:

- Forestry compensation measures (afforestation with arolla pine and larch) to positively impact species composition and ecological conditions.
- Development of two wetlands with appropriate plants at the margins to improve the ecosystem.
- Habitat improvement in areas where gravel is removed (see Fig. 7).
- Revitalization of an ancient peat bog as well as a moor which has been degraded due to drainage. Through irrigation and the removal of trees, the offset targets will have positive effects on the ecosystem and the landscape by developing typical open landscapes for high moors.

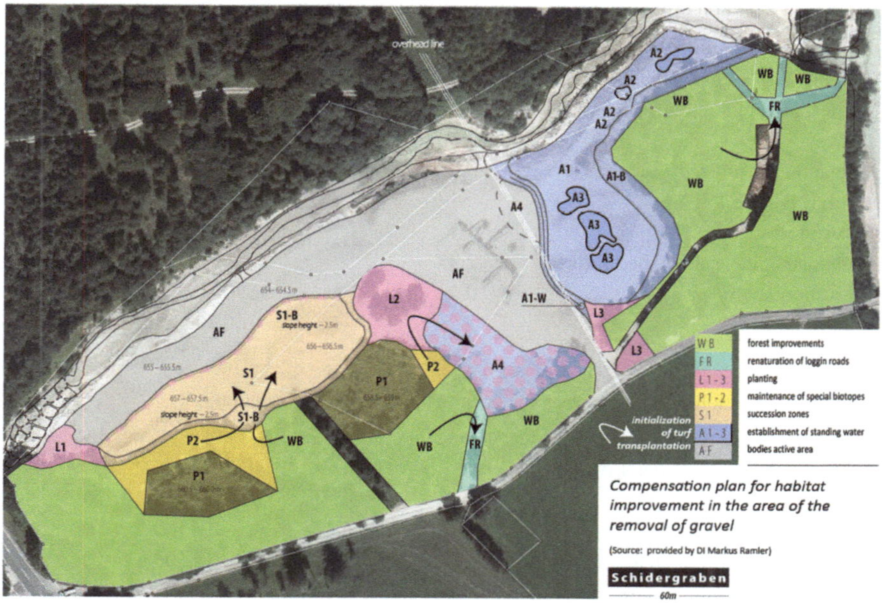

Fig. 7 Compensation plan for habitat improvement in the area of gravel extraction (Source: plan provided by DI Markus Ramler including own revisions, orthophoto: Salzburg State Government, digital map: BEV, planning of measures: ENNACON KG in co-operation with ÖBf)

2.2 *Evaluation of Impacts on Nature and Landscape as well as Determination of Offsets*

In the evaluation of impacts and offsets, the project applicants and auditors together discussed and agreed on a final calculation of the measures based on the guidelines by Loos (2006) (see Sect. 1.2). According to these guidelines, the effects before and after the impacts and compensation are considered, including the duration of the effects. However, the correction factor for the consideration of the time dimension only reflects a maximum of 20 years, whereas the project encompasses a time-frame of over 100 years. To take sufficient account of the duration of the impact effects of the measures, the impacts are split into five phases of up to 20 years each, with the impacts on landscape and the ecosystem calculated separately for each phase and affected area. To illustrate the evaluation of the impacts, we present the calculation as stated in the UVG for phase 1 and quarrying area 1 covering a time-frame of 18 years (see Tables 6, 7 and 8).

The evaluation related to the ecosystem is based on the value scores according to different biotopes and land use types (see also Table 3). Related to the landscape, the value score for the different types of landscapes (see Sect. 1.2) and the supplement factor for evaluating impacts on the recreational value (see Table 5) need to be considered. For the ecosystem and the landscape, the correction factor reflecting the effect duration of 18 years is set to 0.8 (see also Table 4).

Table 6 Evaluation of the impact on the ecosystem before the intervention

Biotope/land use type	Value score (VS)	Area in m^2	Valuation
Existing quarry and company area			
Open spaces (e.g. mining)	0.0	41,200	0
Conveyor line	0.0	23,400	0
Operationally utilized area	0.0	37,300	0
Quarrying area 1			
Spruce forest	1.7	5200	8840
Carbonate, pine, spruce, beech forest	2.3	50,700	116,610
Extraction quarry	0.0	8900	0
Sum		166,700	125,450

Table 7 Evaluation of the impact on the ecosystem after the intervention

Biotope/land use type	Value score ecosystem (VS_E)	Area in m^2	Valuation
Existing quarry and company area			
Open spaces	0.0	6500	0
Conveyor line	0.0	19,100	0
Operationally utilized area	0.0	38,500	0
Recultivation (planting of neophytes)	2.0	3600	7200
Recultivation (greening of rock slopes)	1.3	6200	8060
Recultivation (afforestation)	1.3	28,000	36,400
Quarrying area 1			
Open spaces	0.0	9600	0
Conveyor line	0.0	10,100	0
Recultivation	1.3	2800	3640
Recultivation	1.3	37,900	49,270
Maintenance paths	1.0	1700	1700
Unclaimed forests	2.3	2700	6210
Sum		166,700	112,480

Table 8 Evaluation of the intervention impact on the landscape

Impact affected area in m^2	166,700
Value score landscape (VS_L) before impact	2.3
Value score landscape (VS_L) after impact	−0.4
Supplement factor for recreational value	1.0

Based on the guidelines by Loos (2006), the value scores for impacts on the ecosystem are calculated by subtracting the sum of the evaluation of the impacts on the ecosystem after the intervention (112,480, see Table 7) from the sum of the evaluation of the impacts on the ecosystem before the intervention (125,450, see Table 6). Then the resulting difference (−12.970) is multiplied by the correction factor of the effect duration (0.8) (see Fig. 3, Sect. 1.2 for the calculation method).

Table 9 Evaluation of the impact on the ecosystem before and after compensation (correction factor for timing of compensation implementation and for effect duration for phases 1–3: 1.0)

Biotope/land use type	Value score (VS)	Area in m^2	Valuation
Phase 1 before compensation			
Drainage ditch	2.3	7525	17,308
Severely degraded moor area west	2.7	27,575	74,453
Partly degraded moor area east	3.3	32,555	107,432
Former peat-cutting area	2.0	15,065	30,130
Phase 1 after compensation (year 20)			
Drainage ditch (begin of irrigation)	2.3	7525	17,308
Severely degraded moor area west (felling activities)	3.0	27,575	82,725
Partly degraded moor area east	3.3	32,555	107,432
Former peat-cutting area (felling activities)	2.7	15,065	40,676
Value score compensation ecosystem phase 1 (VS $Compensation_E$ 1): 18,818			
Phase 2 after compensation (year 50)			
Drainage ditch (irrigation)	2.7	7525	20,318
Severely degraded moor area west (regeneration)	3.3	27,575	90,998
Regenerated moor area east	3.7	32,555	120,454
Former peat-cutting area (renaturation)	3.3	15,065	49,715
Value score compensation ecosystem phase 2 (VS $Compensation_E$ 2): 33,344			
Phase 3 after compensation (year 100)			
Drainage ditch (irrigation)	3.3	7525	24,833
Regenerated moor area west	4.0	27,575	110,300
Regenerated moor area east	4.0	32,555	130,220
Former restored area	3.7	15,065	55,741
Value score compensation ecosystem phase 3 (VS $Compensation_E$ 3): 44,130			

This gives a total value score of impacts on the ecosystem VS $Impact_E = -10,376$ for phase 1. To calculate impacts due to the intervention in the landscape, the impact affected area, the values score of the landscape before and after the impact, the supplement factor for the recreation value (see Table 8) and the correction factor of the effect duration (0.8) are multiplied. Here the value score of impact on the landscape VS $Impact_L = -122,691$ for phase 1 (see Fig. 4, Sect. 1.2 for the calculation method). Summing up, the VS $Impact_L$ and VS $Impact_E$ gives a total score for the impacts of −133,067 for phase 1. The impact for all five phases was determined as −964,520. Here +168,020 credit points were taken into account from phases 3 to 5 due to positive partial results in the course of large-area recultivation measures and increasing ecosystem quality during the last phases.

To evaluate the impact of the compensation on the ecosystem and landscape before and after implementation, the offsets are calculated separately and, if necessary, in several phases. To exemplify the calculation of offset measures, we present the evaluation of the revitalization of the moor. This revitalization process takes place in three phases and in different parts of the moor (see Tables 9 and 10). The

Table 10 Evaluation of the impact of compensation on the landscape (correction factor for timing of compensation implementation and for effect duration for phases 1–3: 1.0)

Phase 1	
Area affected by impact in m^2	82,720
Value score landscape (VS_L) before compensation	3.3
Impact factor landscape	0.4
Value score compensation landscape phase 1 (VS $Compensation_L$ 1): 109,190	
Phase 2	
Area affected by impact in m^2	82,720
Value score landscape (VS_L) before compensation	3.7
Impact factor landscape	0.2
Value score compensation landscape phase 2 (VS $Compensation_L$ 2): 61,213	
Phase 3	
Area affected by impact in m^2	82,720
Value score landscape (VS_L) before compensation	4.0
Impact factor landscape	0.2
Value score compensation landscape phase 3 (VS $Compensation_L$ 3): 66,176	

different areas are given individual value scores depending on their ecological quality. The value scores for the ecosystem and landscape increase over the phases due to an increase in quality and affected areas. The correction factors for the timing of compensation implementation and effect duration are stable. Based on the values stated in Tables 9 and 10, a total value score of +332,870 is achieved for the moor revitalization. Together with the forestry compensation measures, the development of two wetlands and habitat improvements, the compensation measures for the quarry expansion achieve a total compensation score of +804,823.

Table 11 confirms that, according to the calculation of impacts and compensation, the suggested offset measures will contribute to an improvement of the ecosystem and landscape, and thus the project could be approved. The offsets also meet the criteria of spatial proximity to the impacts. Based on the UVE and UVG, the responsible authority approved the project, stating in the notification based on the UVP-G 2000 that the project is of public interest, no adequate alternative solutions exist, and thus, in accordance with § 3a para. 2 Sbg NSchG, the project can be realized under the condition that the compensation measures suggested by the project applicants and auditors are implemented. To ensure the adequate realization of the offsets, a monetary security has to be deposited by the project applicants (see also Sect. 2). The amount of the security reflects the expense of recultivation measures for up to 3 years. The deposit will be released when the ecological building supervision confirms in an evaluation report released between 3 and 5 years after project end that all compensation measures have been appropriately realized.

Table 11 Total value score of impacts and compensation for the quarry extension

	Value scores
Impact	−964,520
Credits	+168,020
Compensation	+804,823
Balance	+8,323

3 Conclusions from Austrian Practitioners Regarding an EU-Wide No Net Loss and Offset Strategy

According to local practitioners, although initial experiences in the implementation of impact mitigation regulations have been made in Austria, more efforts are necessary to implement an impact mitigation regulation of high quality to contribute to an efficient NNL. Thus, according to the findings of an expert panel on impact mitigation in Austria organized by the Austrian Federal Forestry Office (Österreichische Bundesforste 2014) and according to an interview with Dr. Wolfgang Wiener, the Environmental Ombudsman in Salzburg (see box at the end of this section), a range of challenges have to be considered in this regard: (1) implementation of standard legal requirements, (2) establishment of transparent and objective offset metrics, (3) provision of suitable compensation areas and monitoring of offsets and (4) prevention of weakened conservation categories and setting of "no-go areas" for impacts.

According to European experts, one major challenge to an efficient European-wide NNL is the lack of mandatory approaches to offset regulations at local scales (Tucker et al. 2013). Austrian practitioners concur with this opinion. As demonstrated in Sect. 1.1, the Austrian states differ substantially in their legal implementation and formation of impact mitigation regulations. Thus, Austrian experts have called for a nationwide legal framework (Bergthaler and Sladek 2014; Proksch and Graf 2011; Ragger et al. 2016). According to Dr. Wiener, a lack of standard, binding regulations can hamper efficient offset measures at the European and national level in the course of cross-state and cross-national projects such as infrastructure and electricity projects.

In addition to mandatory agreed regulations, Austrian practitioners and experts highlight the need for standardized offset metrics to improve the transparency and objectivity of compensation measures (Bergthaler and Sladek 2014; Bußjäger 2013; Jungmeier et al. 2010; Knollconsult Umweltplanung et al. 2016; Österreichischer Rechnungshof 2008; Ragger et al. 2016; Roskosova 2014). This also reflects discussions at the European level (Rayment et al. 2014). The environmental ombudsman in Salzburg has pointed out that, due to the range of different offset regulations in Austria as well as the lack of guidelines on metrics, the determination and design of compensation measures can vary considerably between projects. In particular, the realization of offsets depends on the actors and the negotiation process between the authorities, planning practitioners, experts and project applicants (see also Roskosova 2014; s.n. 2014). According to Dr. Wiener and Bergthaler and

Sladek (2014), a standardized method is also vital to improve the legal framework for compensation measures as well as their acceptance by project applicants. To reduce the criticism of offsets that they are inconsistently applied and that nature conservation can be "bought", the Salzburg Ombuds office has fostered the development of objective metrics to assess intervention in and compensation for nature and landscape, based on the guidelines by Loos (2006) and put into practice by Salzburg's federal authority (see Sect. 1.2). However, Dr. Wiener questions the need for European-wide standardized metrics for biodiversity offsets. Although seeing a strong requirement for transparent and objective offset metrics in European member countries, he emphasizes that there will always be a need for some flexibility to include expert judgments and stakeholder perspectives when negotiating project details and offset realizations (see also Rayment et al. 2014 for a European perspective). He adds that offsets will always depend on the local requirements of the affected area, their surroundings and the availability of compensatory land at a reasonable price, all of which factors can differ significantly between countries and regions.

According to one member of the expert panel organized by the Austrian Federal Forestry Office (s.n. 2014), a mitigation bank can circumvent the problem of land prices shooting up as soon as areas are designated for compensation. To foster the establishment of mitigation banks for potential offsets, which do not yet exist in Austria (see Sect. 2), the Austrian Federal Forestry Office plans to provide a service for the management of natural areas including the planning and implementation of offsets, provisioning of suitable compensation areas and the monitoring of implementation quality. Dr. Wiener points out that there is a particular requirement for long-term monitoring and ongoing maintenance of compensation measures to secure an effective offset initiative (see also s.n. 2014).

According to European experts for an EU-wide NNL, competent monitoring is crucial to prevent the weakening of environmental and nature conservation standards by offsets (Tucker et al. 2013). Furthermore, Dr. Wiener emphasizes that, in the course of a European NNL, it is of paramount importance that standards for protecting species and habitats not be weakened under the umbrella of biodiversity offsets. He suggests the implementation of European "no-go areas" which cannot be developed due to their ecological and environmental value. Thus, an effective NNL at the European level needs to secure correct implementation of the mitigation hierarchy to avoid offsets replacing the objectives of avoidance and minimization, and project developers skipping straight to compensation (Tucker et al. 2013).

The biologist **Dr. Wolfgang Wiener** has been Environmental Ombudsman since 1998. He is the head of the Salzburg Ombuds office. His responsibilities are for water protection, fisheries, agriculture, transport, energy and public relations.

(continued)

Environmental authorities (German: *Umweltanwaltschaft*) and their environmental ombudsmen are anchored in each Austrian federal state. They represent the public interests of nature and environmental protection and support citizens in environmental issues (http://www.umweltanwaltschaft.gv.at/de/).

References

Baumgartner C, Eberhartinger-Tafill S (2006a) Leitfaden UVP für Handels- und Freizeiteinrichtungen, Industrie- und Gewerbeparks. Umweltverträglichkeitserklärung. Einzelfallprüfung, Bundesministerium für Land- und Forstwirtschaft, Umwelt und Wasserwirtschaft, Vienna

Baumgartner C, Eberhartinger-Tafill S (2006b) Leitfaden UVP für Schigebiete. Umweltverträglichkeitserklärung. Einzelfallprüfung. Bundesministerium für Land- und Forstwirtschaft, Umwelt und Wasserwirtschaft, Vienna

Bergthaler W (2013) Ausgleichsflächen: Herausforderungen für eine erfolgreiche Umsetzung. Rechtliche Situation in Österreich. ExpertInnenforum Purkersdorf. 27.11.2013. Presentation. www.bundesforste.at/fileadmin/naturraummanagement/naturraummanagement/pdf_expertinnen forum/Ausgleichmassnahmen/01_ExpertInnenforum2013_Bergthaler.pdf. Accessed 8 July 2015

Bergthaler W, Sladek B (2014) Ausgleichsmaßnahmen & Ersatzleistungen: Überblick und Ausblick. NaturRaumManagement 1(19):4–5

BMLFUW (ed) (2009) Indikatoren-Bericht zur Biodiversität in Österreich. BMLFUW, Vienna

BMLFUW (ed) (2013) Indikatoren-Bericht für das Monitoring Nachhaltiger Entwicklung (MONE) 2013. BMLFUW, Vienna

BMLFUW (ed) (2014) Biodiversitäts-Strategie Österreich 2020+. Vielfalt erhalten – Lebensqualität und Wohlstand für uns und zukünftige Generationen sichern! Umweltbundesamt, Vienna

Büchele KT, Eberhartinger-Tafill S (2006) Leitfaden UVP für Intensivtierhaltungen. Umweltverträglichkeitserklärung. Einzelfallprüfung. Bundesministerium für Land- und Forstwirtschaft, Umwelt und Wasserwirtschaft, Vienna

Bußjäger P (2013) Modell(e) der Eingriffsregelung in Österreich. In: Czybulka D (ed) 35 Jahre Eingriffsregelung. Nomos Verlagsgesellschaft, Baden-Baden, pp 165–176

Darbi M, Ohlenburg H, Herberg A, Wende W (2010) Impact mitigation and biodiversity offsets—compensation approaches from around the world. A Study on the application of Article 14 of the CBD (Convention on Biological Diversity). Bundesamt für Naturschutz, Bonn

data.gv.at (2015) Katalog. Gemeindegrenzen Österreich. https://www.data.gv.at/katalog/dataset/c33d36b0-f184-4f2a-89cc-839ca7fcf88a. Accessed 16 Apr 2016

Fiala I (2014) Veröffentlichung zu Zustand und Bedeutung der biologischen Vielfalt in Österreich. In: Online-Fachzeitschrift Ländlicher Raum 01:1–6. www.bmlfuw.gv.at/dms/lmat/land/laendl_entwicklung/Online-Fachzeitschrift-Laendlicher-Raum/Biodiversitaet/06_Fiala_Biodiversit-t/06_Fiala_Biodiversit%C3%A4t.pdf. Accessed 6 July 2015

FSV (2015) Tätigkeitsbericht der Arbeitsausschüsse 2014/2015. FSV, Vienna

Jungmeier M, Stallegger M, Proschek-Hauptmann M (2010) ISOnatura – Qualität von Naturschutzverfahren. Umweltdachverband, Vienna

Klaffl I, Bergthaler W, Niederhuber M, Brandl K, Leitner M, Lexer W, Kurzweil A, Margelik E, Nagl C, Tulipan M (2006) UVP-Evaluation. Evaluation der Umweltverträglichkeitsprüfung in Österreich. Umweltbundesamt, Vienna

Knollconsult Umweltplanung, REVITAl Integrative Naturraumplanung, Haslinger/Nagele & Partner Rechtsanwälte (2016) Ausgleich für Eingriffe in Natur und Landschaft. Studie im Auftrag der Umweltanwaltschaften Burgenland, Niederösterreich, Oberösterreich. http://www.ooe-umweltanwaltschaft.at/xbcr/SID-914F451C-4CC15FBB/StudieEingriffAusgleich.pdf. Accessed 7 Apr 2017

Land of Upper Austria (2012) Bericht des Ausschusses für Wohnbau, Natur- und Landschaftsschutz betreffend das Landesgesetz, mit dem das Oö. Natur- und Landschaftsschutzgesetz 2001geändert wird (Oö. Natur- und Landschaftsschutzgesetz-Novelle 2014). www2.land-oberoesterreich.gv.at/internetltgbeilagen/Beilage%201051/2014%20-%20Ausschussbericht.pdf?id=5436&n=1051&j=2014. Accessed 17 July 2015

Landes-Rechnungshof Vorarlberg (2008) Prüfbericht über den Naturschutzfonds. Bregenz. www.lrh-v.at/wp-content/uploads/2013/04/3090.B.Endbericht.pdf. Accessed 23 July 2015

Loos E (2006) Richtlinie zur Erstellung naturschutzfachlicher Gutachten im Hinblick auf die Bewertung von Ersatz- und Ausgleichsmaßnahmen nach dem Salzburger Naturschutzgesetz. Amt der Salzburger Landesregierung, Salzburg

Österreichische Bundesforste (ed) (2014) Planung – Umsetzung – Betreuung. Ausgleich & Ersatz in der Natur. Natur.Raum.Management 1(19):1

Österreichischer Rechnungshof (2008) Maßnahmen zum Schutz der Umwelt und der Gesundheit im Straßenbau in Österreich. BUND 5:39–66

Peters W, Siewert W, Szaramowicz M (2002) Folgenbewältigung von Eingriffen im internationalen Vergleich. BfN-Skripten 82. Bundesamt für Naturschutz, Bonn

Plattner G (2014) Vermeiden vor Verringern vor Kompensieren. NaturRaumManagement 1(19):3

Proksch T, Graf K (2011) Landschaftskonto. Endbericht. Etablierung des Landschaftskontos als strategischen Modellansatz am Sektor Landschaftsentwicklung im Ballungsraum Wien—Niederösterreich. Report on behalf of the Department Regional Planning and Regional Policy of the Lower-Austrian Government. Vienna

Ragger C, Pröbstl-Haider U, Knoll T (2016) Konzeption für eine Eingriffsregelung in Österreich. Eine freiwillige anzuwendende Arbeitshilfe für die Praxis. Naturschutz und Landschaftsplanung 48(10):305–312

Rayment M, Haines R, McNeil D, Conway M, Tucker G, Underwood E (2014) Study on specific design elements of biodiversity offsets: biodiversity metrics and mechanisms for securing long term conservation benefits. DG Environment, ICF, London

Regionalmanagement OÖ (2015) Ausschreibung für LIFE-Projekte, 18 June 2015. www.rmooe.at/aktuelles/ausschreibung-f%C3%BCr-life-projekte. Accessed 15 July 2015

Roskosova J (2014) Landschaftspflegerische Kompensationsmaßnahmen im Straßenbau. Planung, Umsetzung, Erhaltung. Dargestellt an realisierten Abschnitten der S1 – Wiener Außenring Schnellstraße. AV Akademikerverlag, Saarbrücken

s.n. (2010) Tirol überlegt die "Ausgleichsregelung" einzuführen. Landesumweltanwaltschaft berichtete bei einer Enquete über die Erfahrungen in Salzburg. In: LUA-Notizen 3

s.n. (2014) Verhandlungsmasse Natur? ExpertInnenforum "Ausgleichsflächen". NaturRaum Management 1(19):6–7

Schiffner W, Matzinger A (2015) Das oberösterreichische Naturschutzrecht. Das oö. Natur- und Landschaftsschutzgesetz 2001 samt Kommentar, Durchführungsverordnungen und weiteren Rechtsgrundlagen. Schriftenreihe des Landes Oberösterreich, Band 13. Amt der Oö. Landesregierung, Linz

Sladek B (2013) Aus der österreichischen Praxis. ExpertInnenforum Purkersdorf. 27.11.2013. Presentation. www.bundesforste.at/fileadmin/naturraummanagement/naturraummanagement/pdf_expertinnenforum/Ausgleichmassnahmen/02_ExpertInnenforum2013_Sladek.pdf. Accessed 8 July 2015

Statistik Austria (2015a) Bevölkerungszahl Österreichs stieg auf fast 8,58 Mio. zu Jahresbeginn 2015. Pressemitteilung: 10.972-032/15 (last update 11.06.2015). www.statistik.at/web_de/statistiken/menschen_und_gesellschaft/bevoelkerung/080912.html. Accessed 6 July 2015

Statistik Austria (2015b) Bevölkerung (last update 16.06.2015). www.statistik.at/web_de/statistiken/menschen_und_gesellschaft/bevoelkerung/index.html. Accessed 6 Jul 2015

Süssenbacher R, Jungmeier M (2017) Ausgleichsmaßnahmen im Naturschutzrecht: das Beispiel Österreich. Naturschutz und Landschaftsplanung 49(10):317–324

Tucker G, Allen B, Conway M, Dickie I, Hart K, Rayment M, Schulp C, van Teeffelen A (2013) Policy options for an EU no net loss initiative. Report to the European Commission. Institute for European Environmental Policy, London

Umweltbundesamt (ed) (2012) UVE-Leitfaden. Eine Information zur Umweltverträglichkeitserklärung, Überarbeitete Fassung. Umweltbundesamt, Vienna

Umweltbundesamt (2015a) Nationalparks, Naturschutzgebiete & Co. www.umweltbundesamt.at/umweltsituation/naturschutz/sg/. Accessed 6 July 2015

Umweltbundesamt (2015b) Überblick zum UVP-G 2000. www.umweltbundesamt.at/umweltsituation/uvpsup/uvpoesterreich1/uvpg/. Accessed 9 July 2015

Umweltbundesamt (2015c) Anzahl und Art der Genehmigungsverfahren. http://www.umweltbundesamt.at/umweltsituation/uvpsup/uvpoesterreich1/verfahrensmonitoring/gv_artzahl/. Accessed 10 July 2015

Umweltbundesamt (2015d) Online-Abfrage UVP-Genehmigungsverfahren. www.umweltbundesamt.at/umweltsituation/uvpsup/uvpoesterreich1/uvpdatenbank/uvp_online/. Accessed 10 July 2015

Belgium

Valérie Dupont, Hendrik Schoukens, and Charles-Hubert Born

There has been an alarming decline in biodiversity in recent decades, described by some as the "sixth extinction" in earth's long history.[1] Similarly, the innumerable services provided by ecosystems are under pressure. These will be very expensive—if not impossible—to replace by technological solutions.[2] The main cause of this in Europe is rooted in land use changes and the intensive exploitation of natural resources.[3] The primary goal of the European Union's Biodiversity Strategy to 2020 is to halt the loss of biodiversity and ecosystem services.[4] In order to reach this target, it is essential that the European Union and its member states explore innovative solutions, as conventional methods for the prevention and remediation of environmental damage (regulation, licensing, protected areas, environmental impact studies, etc.) appear insufficient to

[1]Richard E. Leakey and Roger Lewin, *The Sixth Extinction: Patterns of Life and the Future of Humankind* (Doubleday New York 1995). See more recently: Anthony D. Barnosky et al., 'Has the Earth's sixth mass extinction already arrived?' (2011) Science 51.

[2]Millennium Ecosystem Assessment, *Ecosystems and Human Well-Being: Biodiversity Synthesis* (2005).

[3]European Environment Agency (EEA), *The European Environment—State and Outlook 2015* (2015).

[4]European Commission, *Communication from the Commission to the European Parliament, the Council, the Economic and Social Committee and the Committee of the Regions, Our life insurance, our natural capital: an EU biodiversity strategy to 2020* (COM(2011) 244 final, 2011) (hereinafter 'EU Biodiversity Strategy'). The Biodiversity Strategy was endorsed by the Council of the European Union in its Decision of 21 June 2011 (EU Biodiversity Strategy to 2020—Council conclusions, 11978/11).

V. Dupont (✉) · C.-H. Born
University of Louvain, Louvain-la-Neuve, Belgium
e-mail: v.dupont@uclouvain.be; charles-hubert.born@uclouvain.be

H. Schoukens
Ghent University, Ghent, Belgium
e-mail: Hendrik.Schoukens@UGent.be

W. Wende et al. (eds.), *Biodiversity Offsets*,
https://doi.org/10.1007/978-3-319-72581-9_5

halt the decline.[5] As part of these innovations, Action 7b of the strategy required that the European Commission proposed a No Net Loss Initiative by 2015 by implementing, for example, compensation and offset schemes aiming to counterbalance development impacts.[6]

At present, no comprehensive and coherent framework exists for the design and implementation of biodiversity offsets at EU level. However, the absence of a legally binding No Net Loss scheme at EU level should not lead us to negate the compensation requirements present in several existing EU environmental directives. Several regulatory instruments provide important incentives and, in some instances, mandatory requirements in terms of biodiversity offsetting. The Habitats Directive,[7] the Birds Directive,[8] the Strategic and Environmental Impact Assessment Directives[9] and the Environmental Liability Directive[10] (ELD) all require competent authorities to consider, to varying degrees, compensation or offsets for plans, projects (or accidents) that will (or have) significantly impact(ed) the environment.

Yet there are substantial differences in the exact scope and binding nature of such compensation or offsetting, and whether they result in a no net loss-obligation. As early as 1985, the Environmental Impact Assessment Directive required that EIAs include a description of measures envisaged in order to avoid, reduce, and, if possible, remedy/offset significant adverse effects.[11] However, since the outcome of an EIA is not considered legally binding by the permit issuing authority, it has often not proved to be a genuine trigger for robust mitigation and offsetting programmes on the ground. By contrast, the outcome of an 'appropriate assessment'

---

[5]EFTEC and IEEP, *The Use of Market-Based Instruments for Biodiversity Protection – the Case of Habitat Banking, Technical Report for European Commission DG Environnement* (2010), 16.

[6]For a definition of biodiversity offsets, see chapter "Introduction of a European Strategy on No Net Loss of Biodiversity" of this book.

[7]Directive 92/43/EEC of 21 May 1992 on the conservation of natural habitats and of wild fauna and flora (1992) OJ L206/7 (hereinafter 'Habitats Directive').

[8]Directive 2009/147/EC of the European Parliament and of the Council of 30 November 2009 on the conservation of wild birds (2010) OJ L20/7, replacing the original Birds Directive, Directive 79/409/EEC of 2 April 1979 on the conservation of wild birds (1979) OJ L103/1 (hereinafter 'Birds Directive').

[9]Directive 2011/92/EU of the European Parliament and of the Council of 13 December 2011 on the assessment of the effects of certain public and private projects on the environment (2011) OJ L26/1 as amended, replacing the original EIA directive, Council Directive 85/337/EEC of 27 June 1985 on the assessment of the effects of certain public and private projects on the environment (1985) OJ L175/40 (hereinafter 'EIA Directive'); Directive 2001/42/EC of the European Parliament and of the Council of 27 June 2001 on the assessment of the effects of certain plans and programmes on the environment (2001) OJ L197/30 (hereinafter 'SEA Directive').

[10]Directive 2004/35/EC of the European Parliament and of the Council of 21 April 2004 on environmental liability with regard to the prevention and remedying of environmental damage (2004) OJ L143 (hereinafter 'ELD').

[11]See Repealed Council Directive 85/337/EEC on the assessment of the effects of certain public and private projects on the environment, art 5.2 and annex IV.

under Article 6(3) of the Habitats Directive is to be considered instructive for the further outcome of the decision-making process for potentially harmful plans and projects in the context of the EU-wide Natura 2000-Network. As a matter of principle, plans or projects that substantially damage any Natura 2000 site cannot be authorized unless application has been made of the derogatory clause. Furthermore, Article 6(4) of the Habitats Directive requires that compensatory measures be taken to ensure the overall coherence of the network if competent authorities decide to approve a project or plan despites its negative impact on a Natura 2000 site.[12] Likewise, compensatory measures are to be considered in the context of project developments that interfere with the framework of the species protection regime in order to ensure that derogations do not significantly affect the survival of the species at stake.[13] In the context of the remediation of environmental damage under the ELD, compensatory measures must be included if the primary remediation does not result in full restoration of the damaged natural resources or services.[14]

The regulatory framework pertaining to biodiversity offsets in Belgium is poorly developed and largely confined to implementing the above-mentioned EU-rules, often with very limited additional guidance. However, for specific types of biodiversity, such as permanent grasslands, wetlands and forests (woodlands), additional mitigation requirements that exceed the EU minimum standards are applied.

Due to the country's federal structure and distribution of powers,[15] existing offset schemes are mostly implemented at the regional level. Various constitutional reforms have transferred many competences to the country's three regions and communities, while reserving certain areas to the federal government. Article 6(1)(II)(1) of the 1980 Special Act on Institutional Reform granted the regions exclusive power over most environmental matters, including nature conservation and permitting schemes relating to environmental impacts.[16] In this context, the regulatory

[12]Habitats Directive 92/43/EEC, art 6(4).

[13]European Commission, *Guidance Document on the Strict Protection of Animal Species of Community Interest under the Habitats Directive 92/43/EEC* (2007), 63.

[14]See ELD, Annex II. This type of remediation is referred to as "complementary" in the ELD. Such "compensatory remediation" aims to compensate for interim losses of natural resources and/or services that arise from the date of damage until primary remediation has achieved its full effect. See Pascale Steichen, 'Le principe de compensation, un nouveau principe du droit de l'environnement?' in *La responsabilité environnementale, prévention, imputation, réparation* (2009), 156–159.

[15]The federal structure of Belgium is rather complex. Belgium is a federal state, comprised of three communities and three regions, each one enjoying exclusive and autonomous power in its material sphere of competences. The two types of federated entities have different sphere of competence but overlapping territories. The federal government, on the other hand, exercises its power over the whole territory of Belgium. Each federal entity is composed of a legislative and an executive branch; the judiciary has been maintained, with some minor exceptions, at the federal level. For more information, see André Alen and others, *Treatise on Belgian Constitutional Law* (Kluwer Law and Taxation Publishers 1992); Hubert Bocken and Walter De Bondt, *Introduction to Belgian Law* (Kluwer Law International 2001).

[16]Special Act of 8 August 1980 on institutional reform, Belgian Official Gazette 15 August 1980 (hereinafter 'Special Act on Institutional Reform'), art 6, § 1, subparas II and III.

framework regarding terrestrial biodiversity offsets differs from region to region. The statutory rules regarding marine protection, including some basic notions of biodiversity offsetting, remain a federal matter of competence and are not treated in this chapter.

1 Theory in Belgium

1.1 Legal Background

As demonstrated below, most provisions requiring or allowing the use of biodiversity offsets in Belgium are determined by regional nature conservation laws. However, their application on the ground is linked to the decisions taken by competent authorities through spatial planning instruments and permit issuance schemes that are integrated, on the one hand, into the planning and building codes, and, on the other hand, into the regulations concerning industrial installations. Thus, these sectoral codes and regulations, combined with the horizontal legislation on environmental impact assessment, determine the scope of the offsetting rules, the type of impacts and activities that are regulated, and the possibilities for the public to participate in the decision-making. They also determine the public agencies that must be consulted, the power of competent authorities to impose certain types of measures and their abilities to monitor their implementation.

Although it would go beyond the scope of this chapter to discuss the distinct legal regimes that govern land use and industrial installations, it is worth mentioning that different types of authorization schemes exist in Belgium's regional regulatory frameworks. On the one hand, a prior environmental permit must be issued for potentially harmful activities, such as the operation of a dairy farm, an airport or an industrial plant. On the other hand, a prior building permit (development consent) must be sought in order to construct such facilities and, more generally speaking, for a wide array of projects that modify the nature or use of land, or interventions such as forest clearing (deforestation) and logging. Most regions have integrated these two permit schemes (mixed permits), or will do so in the near future (e.g. Flemish Region from 2017 onwards). In addition, specific protection schemes are in place for activities susceptible to damage strictly protected species or habitats. Until recently, the species protection regime was not fully integrated into the planning and development control process. It establishes prohibitions and requires a stand-alone derogation scheme. In the Flemish Region, this derogation scheme has been integrated in the planning code since 2009.

1.1.1 Biodiversity Offsets Mechanisms Within the Scope of EU Law

In this section we focus on the offsetting rules pertaining to the Natura 2000 Network, the strict rules governing species protection, and the rules concerning strategic and environmental impact assessments.

Firstly, we will discuss the protection rules that apply to the Natura 2000 sites designated in Belgium by the three regions. Article 6(4) of the Habitats Directive stipulates that public authorities must impose compensatory measures when authorizing a harmful plan or project justified for so-called imperative reasons of overriding public interest. Other than specifying that the compensatory measures must ensure the overall coherence of Natura 2000, the Habitats Directive is relatively silent on the exact content of these measures. The Walloon Region has almost literally transposed this obligation in the Nature Conservation Act without further specifying the meaning of 'compensation' and the terms of its implementation.[17] As of 2016, no permit has been issued in accordance with the Natura 2000 derogation procedure in the Walloon Region.[18] Similarly, the Flemish Nature Conservation Decree[19] does not lay down comprehensive rules pertaining to biodiversity offsetting in the context of Natura 2000. However, Article 36ter, § 5 of that Decree stipulates that an area of habitat (or length in case of linear landscape elements such as hedgerows or stretches of rivers) equivalent to the one destroyed must be re-established and/or restored.[20] In the preparatory work to transposing Article 6 of the Habitats Directive in Flanders, it was stated that compensatory measures should be effective before the damage occurs unless the advance compensation is not necessary to ensure the coherence of the network.[21] In Flanders, the derogation procedure has been applied on numerous occasions, especially in the context of the expansion of ports, including the ports of Antwerp (2000–2002) and Zeebrugge (2000–2005).[22]

In recent years, further clarification has been proposed on the specific requirements to be complied with when applying the derogation clause allowed under

---

[17]Act of 12 July 1973 on nature conservation, Belgian Official Gazette 11 September 1973 (hereinafter 'Nature Conservation Act'), art 29, § 2, subparas 4 and 5.

[18]To the knowledge of the authors, no measures have been notified to the Commission in accordance with article 6(4) of the Habitats Directive in the Brussels Region and in the Walloon Region. Interestingly, the recent decision to approve the renewal of kayak renting on the Lesse would have led to the application of the derogation procedure. In 2015, however, this permit was squashed by the Council of State. See Belgian Council of State, 23 February 2015, no. 230.267, *de Limburg Stirum*.

[19]Flemish Decree of 21 October 1997 concerning nature conservation and the natural environment, Belgian Official Gazette, 10 October 1998 (hereinafter 'Flemish Nature Conservation Decree').

[20]Flemish Nature Conservation Decree, art 36ter, § 5, subpara 2, 2°.

[21]Preparatory Works Flemish Parliament 2001–2002, no. 967/1, 38–39. The Flemish Nature Conservation Act was amended by the Flemish Decree of 19 July 2002, Belgian Official Gazette 31 August 2002 to transpose article 6 of the Habitats Directive. For details, see Hendrik Schoukens, 'Région Flamande' in Charles-Hubert Born and Francis Haumont (eds), *Natura 2000 et le juge – Situation en Belgique et dans l'Union européenne/Natura 2000 and the Judge – Situation in Belgium and in the European Union (actes du colloque organisé à Louvain-la-Neuve, le 31 mars 2011)* (Bruylant 2014), 223.

[22]These measures have not been notified to the Commission. It is not clear whether this was linked to the fact that no priority habitats and/or species were involved. As will be shown below, there is a recent tendency to opt for a broad interpretation of the concept of mitigation in the application of 6 (3) of the Habitats Directive (implementation of article 36ter, § 3, of the Flemish Nature Conservation Decree), thus avoiding the notification procedure to the European Commission.

Article 6(4) of the Habitats Directive. The modification in 2012 of the Brussels Nature Conservation Ordinance with regard to the definition, implementation and monitoring of compensatory measures serves as a good illustration thereof.[23] Based on the guidance document on Article 6(4) of the Habitats Directive established by the European Commission in 2007,[24] the Ordinance itself explicitly specifies the purpose and content of the compensatory measures that have to be adopted when a significant impact on a Natura 2000 site is allowed. In order to ensure the network's coherence, these measures must restore the habitats and species negatively affected in comparable proportions in terms of number and status. They must also provide functions comparable to those that had justified the selection criteria of the original site. Finally, they must be located as close as possible to the site concerned on the territory of the Region.[25] Moreover, Article 64, § 4 of the Ordinance empowers the government to take all necessary actions to ensure the effectiveness of the offset. The appropriate assessment and derogation scheme, including biodiversity offsets, has been extended to regional protected areas.[26]

Secondly, in comparison to the considerable attention paid to the protection regime relating to Natura 2000, the strict rules on protected species have been rather neglected. This is understandable if we remember that Article 16 of the Habitats Directive and Article 9 of the Birds Directive do not explicitly refer to compensatory measures. That said, the European Commission has explicitly reasserted the possibility, when allowing detrimental impacts, of ensuring the maintenance of species populations at a favourable conservation status in their natural range through the positive effects of biodiversity offsets.[27] Neither the Walloon nor the Flemish nature conservation laws lays down explicit rules pertaining to the imposition of compensatory measures in the context of species protection. Although not expressly referred to, mitigation or compensatory measures can nevertheless be imposed by the competent authority if a derogation is issued under the Walloon Nature Conservation Act or the Flemish Species Protection Regulation of 2009, in order to ensure that the

---

[23]Brussels Ordinance of 1 March 2012 relative to nature conservation, Belgian Official Gazette 16 March 2012 (hereinafter 'Brussels Nature Conservation Ordinance').

[24]European Commission, *Guidance document on Article 6(4) of the 'Habitats Directive' 92/43/ EEC* (2007).

[25]Brussels Nature Conservation Ordinance, art 64, § 2, subpara 3. See Valérie Dupont and Charles-Hubert Born, 'L'ordonnance du 1er mars 2012 relative à la conservation de la nature: une avancée significative pour la préservation de la biodiversité urbaine et périurbaine à Bruxelles (Partie II)' (2014) Aménagement, Environnement, Urbanisme et Droit Foncier 138, 143.

[26]Brussels Nature Conservation Ordinance, art 65. The extension of biodiversity offsets to protected areas is only applicable if the decree creating the protected area establishes conservation objectives. To date, none of the decrees have included conservation objectives.

[27]European Commission, *Guidance Document on the Strict Protection of Animal Species of Community Interest under the Habitats Directive 92/43/EEC*, 63. See also Hendrik Schoukens and An Cliquet, 'Mitigation and Compensation under EU Nature Conservation Law in the Flemish Region: Beyond the Deadlock for Development Projects?' (2014) 10 Utrecht Law Review 194, 201; Geert Van Hoorick, 'Compensatory Measures in European Nature Conservation Law' (2014) 10 Utrecht Law Review 161, 170–171.

derogation is not detrimental to the maintenance of populations of the species concerned at a favorable conservation status in their natural range.[28] The Brussels Ordinance provides more clues in this respect. Applications for derogations under the Brussels species protection scheme should include a mitigation and/or compensatory programme if it harms the integrity of animal species.[29]

Thirdly, reference needs to be made of the regional rules relating to strategic and environmental impact assessment. Both the EIA and SEA Directives require that environmental impact assessments and strategic environmental assessments include a description of measures envisaged in order to avoid, prevent or reduce and, if possible, offset likely significant adverse effects on the environment, including on fauna and flora.[30] This requirement has been reinforced by the adoption of the review of the EIA Directive in 2014. In itself, this mitigation requirement has been literally transposed into all regional laws in Belgium. However, the different steps of the mitigation hierarchy are not harmonized between the regions and do not expressly require compensatory measures, especially in regard to impacts arising from projects.[31] While the Flemish Region requires a description of the proposed measures to "avoid, reduce, remedy or compensate",[32] the Walloon Region refers to proposed measures "to avoid, reduce, and if possible, remedy",[33] and the Brussels legislation refers to proposed measures to "avoid, remove, or reduce".[34] Given the terminology used, it is not clear if biodiversity offsets must be proposed to compensate for any residual impacts on biodiversity.

Moreover, as stated above, SEA/EIA procedures have no binding effect on the outcome of the decision-making procedure but leave some discretion to the permit issuing authority. For this reason, even if identified in environmental reports,

---

[28]Walloon Nature Conservation Act, arts 5 and 5bis; Flemish Regulation of 15 May 2009 on the protection and management of species, Belgian Official Gazette 31 August 2009, art 19–23. The Flemish Species Protection Regulation of 2009 also provides the opportunity to frame derogations within the context of so-called species protection programmes, which are aimed at achieving the favourable conservation status for the said species within a given geographical area which corresponds to a so-called 'ecological functional unit' of a certain species (art 26 and 27).

[29]Brussels Nature Conservation Ordinance, art 84, § 1, 2°, g.

[30]EIA Directive, art 5, § 1, and Annex IV, § 7; SEA Directive, Annex I, g (in combination with art 5, § 1).

[31]For plans and programs, the SEA Directive and its transposition into regional laws include an express reference to compensation. See First Book of the Walloon Environmental Code, art D.56, § 3, 7°; Brussels Ordinance of 18 March 2004 on the assessment of the effects of certain plans and programs on the environment, Belgian Official Gazette 30 March 2004, Annex I(g); Flemish Decree of 5 April 1995 containing general provisions concerning environmental policy, Belgian Official Gazette 3 June 1995, art 4.1.1, 7°.

[32]Flemish Decree of 5 April 1995 containing general provisions concerning environmental policy, art 4.3.7, § 1, 2°, (c).

[33]First Book of the Walloon Environmental Code, art D.67, § 3, 3°.

[34]Brussels Ordinance of 5 June 1997 on environmental permits, Belgian Official Gazette 26 June 1997, art 26, 5°; Brussels Land Use Planning Code of 9 April 2004 (CoBAT), Belgian Official Gazette 26 May 2004, art 135, 5°; art 143, 5°.

offsetting measures have often been poorly integrated into planning permits.[35] However, jurisprudence in this regard has tightened in recent years. Permit issuing authorities are required to take into account all the information included in the environmental impact assessment and need to clearly indicate how they have implemented biodiversity offsets in the applicable permit conditions.[36]

In this regard, it is important to highlight the major advances made by the 2014 review of the EIA Directive that had to be transposed into Belgian (regional and federal) environmental law by 16 May 2017. The 2014 amendments brought clarification to the type of measures that have to be proposed, and systematically refer to "offset" as the final step in the mitigation hierarchy. Furthermore, major amendments were made to ensure the effectiveness of mitigation and compensation measures. According to the new Article 8a(1)(b), the decision to grant development consent shall now incorporate any environmental conditions attached to the decision, including measures envisaged to avoid, prevent or reduce and, if possible, offset significant adverse effects on the environment as well as, where appropriate, monitoring measures. In so doing, the directive reinforces the link between the content of the EIA and the decision.[37] Note, however, that nothing prevents competent authorities from requiring different measures to those envisaged by the developer, if more suitable. In addition, member states shall also ensure that the measures envisaged to avoid, prevent or reduce and, if possible, offset significant adverse effects on the environment are implemented by the developer.[38] Last, the 2014 review also provides the opportunity to better integrate biodiversity issues in environmental impact assessments and to coordinate assessments conducted under the Birds and Habitats Directives and the EIA Directive.[39]

To date, no guidelines exist to determine adequate mitigation and compensation measures in the context of the SEA/EIA Directives and its transposition in regional laws, except for wind energy development in Wallonia. This underscores the relative neglect of biodiversity offsetting in relation to SEA/EIA in all the three Belgium's regions. The Walloon Government recently adopted a policy framework specifically for the development of wind energy in which it provides guidelines for the adoption of mitigation and compensation measures. In order to meet its 2020 renewable energy target, the government committed to an objective of generating 3800 GWh

[35]Charles-Hubert Born, Valérie Dupont and Charles Poncelet, 'La Compensation Ecologique des Dommages Causés à la Biodiversité: un Mal Nécessaire?' (2012) Aménagement, Environnement, Urbanisme et Droit Foncier 12, 21.

[36]Belgian Council of State, 27 January 2014, case no 226.219, *Commune de Momignies*. For more details see: Hendrik Schoukens, 'De ondraaglijke lichtheid van milieueffectrapportage: Einmal ist Keinmal?' (2014) Tijdschrift voor Ruimtelijke Ordening en Stedenbouw 21.

[37]Helle Tegner Anker, 'Simplifying Environmental Legislation – Reviewing the EIA Directive?' (2014) Journal for European Environmental & Planning Law 321, 340.

[38]EIA Directive, art 8bis(4).

[39]EIA Directive, art 2(3). Although the Habitats Directive only includes a formal assessment obligation in the context of impacts to Natura 2000 sites, it could be argued that an assessment is also required in order to issue derogations to the strict species protection regime.

through wind turbines on its territory by 2020 and established a reference framework to reach that goal.[40] The goal is to proactively identify suitable areas with regard to wind energy potential as well as territorial, social and environmental constraints. Although this policy framework is not legally binding, it aims to give strategic orientation to developers and permitting authorities for the design and location of windfarms within those suitable areas. The framework identifies potential risks that must be addressed in the environmental impact study at the project development stage, which include impacts on bird species and bats. If, despite mitigation measures and in the absence of alternatives, the proposed project would be liable to cause significant impact on species and naturals habitats protected by European directives, competent authorities may nevertheless approve the project on the condition that compensatory measures be implemented.[41] In the latter case, the environmental impact assessment must identify compensatory measures based on existing studies and guidelines. According to the framework, biodiversity offsets must: be tailored towards the same species and habitats as those impacted by the project, be equivalent, be located as close as possible to the impact, be accompanied by a development plan and be operational at the time of impact. The development application must also include documentation on the legal means through which the compensation site will be made available for the offset actions.[42]

1.1.2 Biodiversity Offsets Beyond the Scope of the European Directives

In addition to the provisions that implement the robust yet patchy EU rules with respect to biodiversity offsetting, some of Belgium's regions have enacted specific rules in order to achieve no-net-loss goals in the context of impacts on "generic" biodiversity. Most notably, the Flemish Nature Conservation Decree provides for several interesting instruments in this respect. It lays down several clauses imposing compensatory measures in relation to activities that impair forms of biodiversity not strictly protected under EU environmental law. However, these clauses are vaguely formulated and, at least until recently, have hardly been applied in practice.

A first interesting instrument is the so-called "duty of care towards nature" ("Natuurzorgplicht") established by Article 14 of the Flemish Nature Conservation Decree, according to which any person involved in the natural environment and conscious of the risk of damage to the natural elements is required to prevent and limit destruction or damage; if this proves impossible then such destruction or

[40]Walloon Government, *Reference framework for the installation of wind turbines in Wallonia* (approved on 21 February 2013) (hereinafter 'Wind Energy Reference Framework'). The reference framework is a non-binding political document intended to apply temporarily until a decretal framework is formally adopted. The initial framework was adopted in 2002. To date, the decree has not yet been adopted.

[41]Wind Energy Reference Framework, 28.

[42]Wind Energy Reference Framework, 28.

damage must be repaired.[43] An accompanying guidance document issued by the Flemish Government back in 1998 emphasized the importance of damage prevention. Damage limitation or compensation is only to be considered in cases where prevention is not feasible. This obligation applies to all individuals, private entities or public authorities, as well as all locations, regardless of the zoning or protection status of the site.[44] However, given the absence of any tangible link to a decision-making framework, this due diligence obligation mostly serves as a fall-back clause to provide some protection in instances where no clear-cut statutory provisions apply. Hence, relatively little biodiversity compensation is explicitly grounded in this duty of care.

Second, the Flemish Nature Conservation Decree contains an overall mitigation scheme for general interventions in nature, even outside sensitive areas. Article 16 of this Decree requires competent authorities to ensure that activities subject to a prior permit or notification cause no avoidable damage to nature. This regulation, termed the "*natuurtoets*", means that the competent authority has to reject an application for an activity resulting in avoidable damage to nature, or impose reasonable measures to avoid, mitigate or—if damage is unavoidable—to compensate the damage. Despite the rather broad scope of the *natuurtoets*,[45] the practical effects remain rather limited. According to the Council of State, the test only applies to "avoidable" damage—to the exclusion of "unavoidable" damage.[46] This implies that a harmful project development cannot simply be rejected whenever the project developer has explicitly demonstrated that he has taken the necessary due caution into account.[47] Hence, according to the letter of the law, there exists no explicit duty to offset "unavoidable" damage as this concept falls outside of the scope of the *natuurtoets*.[48] Permitting authorities do not need to justify the absence of compensatory measures if consideration has been given to the impacts on nature. Clearly, the *natuurtoets* will not guarantee no net loss. The lack of further guidance on how to implement

[43]For more information, see Peter De Smedt and Hendrik Schoukens, 'Natuurzorgplicht: is er leven na het arrest van het Grondwettelijk Hof van 27 mei 2008?' (2009) Niew Juridisch Weekblad 738.

[44]Circular LNW/98/01 of 10 November 1998 relating to nature conservation general measures and on conditions for changing vegetation and small landscape elements according to the regulation of the Flemish Government of 23 July 1998, which lays down detailed rules for the implementation of the Decree of 21 October 1997 concerning nature conservation and the natural environment, Belgian Official Gazette 17 February 1999 (hereinafter '1998 Flemish Nature Conservation Guidance').

[45]Geert Van Hoorick, 'Evolution Récente du Droit de la Conservation de la Nature en Région Flamande' (2014) Aménagement, Environnement, Urbanisme et Droit Foncier 32, 36; Hendrik Schoukens, Karin De Roo and Peter De Smedt, *Handboek Natuurbehoudsrecht*, vol 64 (Kluwer 2011), 110–114.

[46]Belgian Council of State, 7 December 2006, no. 165.664, *bvba Belgicaplant*.

[47]For more information, see Hendrik Schoukens, 'Natuurbescherming Buiten de Lijntjes: de Algemene Natuurtoets als Imperfect Antwoord op het Verlies aan Natuur binnen Vlaanderen' (2014) Tijdschrift voor Omgevingsrecht en Omgevingsbeleid 346.

[48]Van Hoorick, 'Evolution Récente du Droit de la Conservation de la Nature en Région Flamande', 36.

compensation measures in this regard and the absence of strict compliance help to explain the limited effects generated until now by the *natuurtoets*.[49]

Third, all activities which entail the destruction or degradation of vegetation or small-scale landscape elements within protected areas and certain green areas on applicable zoning plans must be expressly authorized.[50] In addition, Article 7, § 1, of the 1998 Flemish Nature Conservation Regulation assigns permanent protection to certain types of threatened habitats (biotopes) such as wetlands, semi-natural grasslands or dunes throughout the whole territory of Flanders.[51] These strict protection rules do not apply when a so-called nature permit has been obtained. If a permit and/or derogation is issued, then the necessary mitigation and compensation measures must be drawn up according to Article 15 § 2, 3° and § 3 of the Flemish Nature Conservation Regulation. Such measures can be proposed by the project proponent or simply imposed in the derogation or permit conditions. At no point has it been explicitly established that a no net loss-policy goal has to be observed in this respect. However, according to the 1998 Flemish Nature Conservation Guidance, permits can only be issued if there is no detrimental effect on nature, which implies that compensatory measures need to be implemented whenever protected habitats disappeared. In light of the lenient approach that has emerged in the case law of the Council of State, ample room is left for qualitative restoration of existing nature values, and there surely exists no absolute ban on impairments to protected biotopes.[52] Given the lack of guidance on how to implement biodiversity offsets in this respect, a so-called baseline loss is not necessarily avoided under the applicable rules, which in many instances are not rigidly applied.[53]

Fourth and most interestingly, the Flemish forest legislation has gradually tried to set out a no net loss policy in the context of forest protection. In 1997 the Flemish Government established a moratorium on deforestation in the Forest Decree.[54] Seen by many as too strict, the Forest Decree was subsequently modified to authorize deforestation in certain circumstances and under strict conditions, requiring compensatory measures in order to maintain the forest cover in the Flemish Region.[55]

---

[49]Schoukens, 'Natuurbescherming Buiten de Lijntjes: de Algemene Natuurtoets als Imperfect Antwoord op het Verlies aan Natuur binnen Vlaanderen' 346.

[50]Flemish Nature Conservation Decree, art 13, § 4.

[51]Regulation of the Flemish Government of 23 July 1998 laying down detailed rules for the implementation of the Decree of 21 October 1997 concerning nature conservation and the natural environment, Belgian Official Gazette 10 September 1998 (hereinafter 'Flemish Nature Conservation Regulation').

[52]See for instance, Belgian Council of State, 20 December 2010, no. 209.868, *Angenon*.

[53]Hendrik Schoukens, Peter Desmedt and An Cliquet, 'The Implementation of the Habitats Directive in Belgium (Flanders): back to the Origin of Species?' (2007) 2 Journal for European Environmental & Planning Law 127.

[54]In 1997, the Flemish Nature Conservation Decree introduced a new article 90bis in the Forest Decree of 1991 that prohibited deforestation unless in pursuit of some public utility work. See Van Hoorick, 'Evolution Récente du Droit de la Conservation de la Nature en Région Flamande', 41.

[55]See ibid, 41.

The authorization scheme is two-tiered. If the deforestation takes place in designated urban zones, is linked to projects of general interest or is part of a nature management plan, the moratorium can be lifted by means of a 'regular' planning permit. All other projects leading to deforestation must be issued with a prior derogation by the competent minister for the environment.[56] If deforestation is allowed, then the clearing must, in principle, be compensated by the permittee.[57] Compensation may be in the form of on the ground outcomes, by paying a sum to a compensation fund, or by a combination of both.[58] However, since 2014, any deforestation bigger than 3 ha must be fully compensated by measurable on the ground outcomes.[59]

In sharp contrast to the Flemish Region, the nature conservation laws of the Walloon Region and Brussels Region do not include additional requirements for compensation for ordinary biodiversity loss. However, the land use planning system in Wallonia establishes a "territorial" compensation mechanism with the potential to offset biodiversity impacts. Any modification of a regional zoning plan ('plan de secteur') to create a new urban zone must be compensated. Two kinds of measures may be considered. First, the creation of a new urban zone can be compensated by the de-urbanisation of some other existing zone, meaning that an equivalent area previously zoned as urban must become rural/non-urban.[60] The idea is to maintain the overall balance (as set in 2005, when this rule entered into force) between urban and rural zones of the concerned land use plan, thereby containing urban sprawl at the regional level. Currently, however, the mechanism has little value for biodiversity conservation. The swapped zones must be equivalent in terms of acreage but do not need to contain equivalent habitats in terms of type, functions and ecological conditions. Second, the Code authorizes an "alternative compensation", which consist of any compensatory measures "defined by the Government in terms of operational, environmental, energetic or mobility terms, taking into account, especially, the impact of the urban zone on the neighborhood".[61] The important leeway granted to the Government to decide the nature and extent of the alternative compensation allows it to impose biodiversity offsets as well as specific operation conditions or gentle mobility schemes for the future urban zone, according to the impact on "the neighborhood". The current government is studying the possibilities of using both kinds of compensation as a tool for biodiversity protection.[62] Interestingly, the Brussels Region

---

[56]Flemish Forest Decree of 13 June 1990, Belgian Official Gazette 28 September 1990 (hereinafter 'Flemish Forest Decree'), article 90bis.

[57]Flemish Forest Decree, art 90bis § 2.

[58]Flemish Forest Decree, art 90bis, § 4.

[59]Flemish Forest Decree, art 90bis, § 1, subpara 2 as modified by the Flemish Decree of 28 February 2014, Belgian Official Gazette 25 March 2014.

[60]Walloon Spatial Development Code (hereinafter CoDT), art D.II.45, § 3, subpara. 1.

[61]CoDT, art D.II.45, § 3, subpara. 1.

[62]Walloon Government, *Network Wallonia Nature, Catalogue of actions, version III* (February 2015), Action 1.B.02. The catalogue of actions is used as a substitute for a real nature conservation strategy at the regional level. In principle, art D46 of the Walloon Environmental Code provides for

has expressed interest in this mechanism as a tool to achieve no net loss of biodiversity in its 2016 nature conservation strategy. The strategy encourages the government to use territorial compensation when creating new urban zones and to push for the inclusion of this requirement in the Brussels Land Use Planning Code (CoBAT).[63]

2 Methodological Background

2.1 *Mitigation Hierarchy*

Most legal provisions mentioned in the previous section envisage the application of biodiversity offsets within the context of the mitigation hierarchy. Compensation is to be applied after appropriate avoidance and minimization measures have been undertaken. For instance, the so-called *natuurtoets* included in Article 16, § 1 of the Flemish Nature Conservation Decree explicitly requires public authorities to impose reasonable measures to prevent, limit, or where unavoidable, to compensate (restore) damage to nature. The European directives and their transposition in national laws also aim to avoid and minimize the impact before taking appropriate measures to compensate the residual impact of the project. However, no guidance exists as to the appropriate level of avoidance and minimization measures. In fact, legal provisions often remain silent in this respect, leaving considerable leeway in evaluating the steps of the hierarchy, especially in the case of the regional offset schemes (Flemish Region). In the context of the mitigation schemes that apply to regional protected biodiversity, judges are already satisfied when they see that offsets have been included in the permit application. For instance, recent case law demonstrates that the requirements of Article 16, § 1 of the Flemish Nature Conservation Decree are considered to be met whenever nature restoration measures have been linked to the projected development. Previously, the *natuurtoets* was often overlooked in the context of permit applications. This also helps to explain the leniency pertaining to the application of the mitigation hierarchy.[64]

As for offset schemes that fall within the ambit of EU nature conservation law, the mitigation hierarchy is established within the assessment and decision-making framework. In the context of the Natura 2000 protection regime, the European Commission says that mitigation measures may be taken into account in assessing whether there is a significant impact on a Natura 2000 site under article 6(3) while compensatory measures

---

the government to adopt a strategy for the conservation of nature. After years of attempting to adopt a nature plan, the current government decided to make an inventory of actions undertaken for the conservation of nature in lieu of a real strategy. The catalogue of actions also contains action to undertake in the future.

[63]Brussels Government, *Brussels Regional Nature Plan 2016–2020* (14 April 2016), Goal 5, Action 5.

[64]Schoukens, 'Natuurbescherming Buiten de Lijntjes: de Algemene Natuurtoets als Imperfect Antwoord op het Verlies aan Natuur binnen Vlaanderen' 346.

should only be considered in the framework of the derogation scheme under article 6 (4).[65] In the same vein, the Brussels Ordinance expressly authorizes competent authorities to take mitigation measures into account when deciding whether the project will have a significant impact on a Natura 2000 site, but prevents them from considering compensatory measures.[66] These cannot be considered as mitigation measures to ensure the integrity of the site and must only be adopted at the derogation stage.[67] Such a stepped approach to the decision-making process in the Natura 2000 protection regime was recently confirmed by the European Court of Justice (see infra).[68] The EIA Directive, as modified by the 2014 review, also distinguishes between measures to avoid and prevent significant impacts from measures to compensate these impacts. In determining whether an EIA should be conducted (screening-stage), competent authorities are to consider measures proposed by the developer to avoid or prevent significant adverse effects on the environment.[69] On the other hand, measures aimed at compensating the impact are only to be considered within environmental impact assessments. The European Commission also adopts a similar position in the framework of species protection. Mitigation measures may be undertaken to ensure the continued ecological functionality of breeding sites or resting places and avoid the strict derogatory regime specified in Article 16(1) of the Habitats Directive.[70] These are different from compensatory measures, which are independent from the project and may only be considered in assessing whether to issue derogations or not.[71]

Clearly, given the above, the qualification of measures as either mitigation or compensation is of the utmost importance. To put it bluntly, the more liberal approach is given to the concept of mitigation, the more leeway is offered to project developers. They are no longer required to frame their damaging projects within the confines of the strict requirements of the derogation clauses. Accordingly, they are no longer obliged to substantiate that the long-term interests tied to their project developments outweigh the damage to EU protected nature which might give way to more leeway for inherently unsustainable project developments which further put in jeopardy existing Natura 2000 sites and protected species. For the European Commission, mitigation measures are an integral part of the specifications of a plan or project. They aim at minimizing or even eliminating the negative impact of a plan or project. Compensatory measures, on the other hand, are basically independent of the plan or project details, and include restoring, creating or enhancing a habitat area or a

[65]European Commission, *Guidance document on Article 6(4) of the 'Habitats Directive' 92/43/EEC*.

[66]Brussels Nature Conservation Ordinance, art 64, § 1 (4); § 2, (1).

[67]Brussels Nature Conservation Ordinance, art 64, § 1, (5).

[68]Case C-521/12 *T. C. Briels and Others* (15 May 2014) (hereinafter '*Briels*'), Case C-387/15 *Hilde Orleans and Others* (21 July 2016) (hereinafter '*Orleans*').

[69]EIA Directive, art 4(4) and (5).

[70]European Commission, *Guidance Document on the Strict Protection of Animal Species of Community Interest under the Habitats Directive 92/43/EEC*, 48.

[71]Ibid, 48.

species population to counterbalance the residual damage, if any, caused by a plan or project.[72] In spite of the importance of strictly delimiting genuine mitigating measures from compensation, no additional specification is given when it comes to the different types of measures. Qualifying a particular measure as either compensation or mitigation is therefore left to the discretion of competent authorities under the supervision of the Council of State. As is demonstrated below, the obfuscation between mitigation and compensation at EU level has trickled down into the planning practices at the regional level in Belgium.

Indeed, developers and competent authorities tend to adopt, in practice, a broad definition of mitigation measures in order to avoid the burdensome derogation procedures. One fear is that such procedures will considerably delay projects and that private projects will never qualify as 'imperative reasons of overriding public interest' making derogations impossible to obtain. Consequently, on-site habitat creation, restoration and enhancement measures are increasingly being integrated into project developments in the hope of facilitating the decision-making procedure. The restoration and/or creation actions aim to ensure that the overall impact ('the ecological footprint') of the project remains below the significant threshold, hence avoiding the derogation procedure set by the Birds and Habitats Directives. Such approaches have engendered additional discussions regarding the distinction between 'mitigation' and 'compensation'.[73]

Likewise, the Walloon Government has recently introduced some 'flexibility' in applying the protection rules by defining broader Natura 2000 site conservation objectives in the regulation establishing conservation objectives for Natura 2000 sites.[74] Conservation objectives are the cornerstone of the Natura 2000 protection regime. According to the European Court of Justice, the likely significant impacts on a Natura 2000 site must be assessed in terms of these objectives.[75] Accordingly, a broad definition of conservation objectives offers greater latitude in assessing projects likely to significantly impact Natura 2000 sites. The regulation establishing the regional and local conservation objectives for species and natural habitats provides that, at the site level, the conservation objectives consist of the "maintenance" of the surfaces of habitats/populations of species and the quality of the natural habitats and species habitats for which the site is designated. From a quantitative perspective, "maintenance" of species and natural habitats may be considered ensured in the case of the displacement of small areas of natural habitats and species habitats for which the site is designated if such displacement is counterbalanced by the prior restoration of an equivalent area in terms of quantity and quality within or near the site in question, in compliance with the

[72]European Commission, *Guidance document on Article 6(4) of the 'Habitats Directive' 92/43/EEC*, 10.

[73]Schoukens and Cliquet, 2014.

[74]Walloon Regulation of 1 December 2016 establishing conservation objectives for Natura 2000 sites, *Belgian Offical Gazette*, 22 December 2016.

[75]Case C-258/11, *Sweetman and others* (11 April 2013), paras 30 & 40.

proportionality principle.[76] This provision, therefore, allows certain interventions in the natural environment if they are limited to habitat 'displacement' or exceptionally, minimal reductions in these surfaces. Based on this provision, an administrative authority can occasionally authorize projects causing damage without invoking the derogation clause if restoration measures are undertaken within the site. As the end result is to meet the conservation objective, the impact is considered insignificant and a derogation is not required.

In the authors' opinion, this flexible approach does not comply with the Directive, in particular given the recent European Court of Justice decision in *Briels* which qualified restoration measures as clear compensatory measures to be undertaken in the framework of Article 6(4). In this case, the Court of Justice was asked to make a decision in the context of a preliminary ruling by the Dutch Council of State as to whether the term "not adversely affecting the integrity of the site" in Article 6(3) of the Habitats Directive is to be interpreted as meaning that, where the project adversely affects the area of a protected natural habitat type within the site, the integrity of the site is not adversely affected if, in the framework of the project, an area of that natural habitat type of equal or similar size is created within the site.[77] The Court concurred with the opinion of the Advocate General who considered that the integrity of a site is indeed affected if a project or plan negatively impacts a natural habitat type present on the site and measures are put in place for the development of the same or a greater area of this habitat type on the site.[78] It follows that such measures should be considered to be a form of compensation. According to the Court, the potential positive effects of the future development of new habitat are difficult to predict and, in any case, will only become apparent after several years.[79] However, based on the precautionary principle, the appropriate assessment under Article 6, paragraph 3 shall not leave any reasonable doubt as to the effects of the proposed works on the protected site concerned.[80]

The strict rationale of the *Briels* decision had earlier been upheld in a case heard by the Belgian Council of State. Although a tendency to opt for a more lenient interpretation of the notion of 'mitigation' is present in all three regions, the Council of State has only had the opportunity to rule on this qualification issue in regard to projects and plans approved in the vicinity of Natura 2000 sites in the Flemish Region. In 2011 the Flemish Government issued a planning permit for the construction of a road bypass in the province of Limburg ('Noordzuidverbinding') which

---

[76] Walloon Regulation of 1 December 2016 establishing conservation objectives for Natura 2000 sites, Article 3, § 1, 1°. For more information, see Valérie Dupont, Hendrik Schoukens and An Cliquet, 'Aspects juridiques de la restauration écologique: un avenir du droit de la conservation de la nature?' (2014) Aménagement, Environnement, Urbanisme et Droit Foncier 84.

[77] Case C-521/12 *T. C. Briels and Others*.

[78] Ibid, para 39.

[79] Ibid, para 32.

[80] For more details see: Hendrik Schoukens, 'Habitat restoration measures as facilitator for economic development within the context of the EU Habitats Directive: balancing no net loss with the prevention principle?' (2017) 29 Journal of Environmental Law 47.

would cut through a Natura 2000 site. Since the creation of a natural corridor zone aimed at addressing the encroachment of the nearby Natura 2000 sites had been included in the project design, the project was eventually authorized without the need to apply the derogation clause of Article 6(4) of the Habitats Directive. In this case, the Belgian Council of State held that the creation of the corridor zone should be labeled a compensatory measure, i.e. it can only be considered when application has been made of the derogation clause of Article 6(4) of the Habitats Directive.[81] At that point, the Belgian Council of State thus overruled the Flemish Government for not observing the mitigation hierarchy regarding Natura 2000 sites.

The previous case is not a stand-alone. In 2012 the Flemish Government authorized a 1000-ha extension of the Port of Antwerp by recognizing the positive effects of a massive nature development plan linked to the harbour extension aimed at the restoration of the affected Natura 2000 sites. However, this integrated planning scheme was also rejected by the Belgian judges since they deemed it incompatible with the strict precautionary approach set out by the EU Nature directives and that it envisaged the destruction of large tracts of protected habitats.[82] Given the impairments to Natura 2000 sites, it could not be maintained that the measures merely aimed to 'prevent' or 'mitigate' damage to nature. By contrast, the project measures seemed to 'offset' the partial destruction of valuable Natura 2000 sites by the creation and restoration of new natural areas in the immediate vicinity of the Port area. In their more recent decisions, the Belgian judges have justified their strict stance by reference to the 2014 *Briels* ruling of the Court of Justice.[83] In its ruling of 21 July 2016, the European Court of Justice has confirmed that measures undertaken to restore habitats before the actual impairment of other patches of habitats has taken place must still be designated as compensation. In other words, such measures are only permissible whenever application has been made of the strict derogation clause (Article 6(4) of the Habitats Directive).[84] This finally prompted the Belgian Council of State to quash the planning permits by its final rulings of 20 December 2016[85] and 12 May 2017.[86] In doing so, the Belgian Council of State steadfastly refuted pleas for a more liberal approach to biodiversity offsetting within the context of Natura 2000 sites. That is not to say that additional restoration cannot be linked to urban planning projects. Yet they cannot be tagged as 'mitigation' and thus serve as justification to bypass the derogation clause in the context of Natura 2000 sites.

As is the case with the Natura 2000 regime, developers increasingly try to avoid the application of the species derogation scheme by incorporating restoration measures into their projects.[87] In Wallonia such practice is common in the context of wind power developments and species protection derogations. Based on the Wind

---

[81]Belgian Council of State, 29 March 2013, no. 223.083, *vzw Natuurpunt Limburg*.

[82]Belgian Council of State, 3 December 2013, no. 225.676, *Hilde Orleans*.

[83]See for instance: Belgian Council of State, 20 November 2015, no. 233.000, *Hilde Orleans*.

[84]Case C-387/15 *Orleans e.a.*, par. 48 and 59.

[85]Belgian Council of State, 20 December 2016, no. 236.837, *Hilde Orleans*.

[86]Belgian Council of State, 12 May 2017, no. 238.186, *Hilde Orleans*.

[87]Schoukens and Cliquet 2014.

Power Reference Framework adopted by the Walloon Government, permits for the installation of wind turbines routinely recognize that compensatory measures planned in the project are relevant to avoid any significant disturbance to birds, in this way avoiding the need for a derogation under article 5 of the Nature Conservation Act.[88] Hitherto, the issue of whether such developments should require a prior derogation has remained unaddressed. Given the very strict reasons specified by Article 9 of the Birds Directive for the issuance of a derogation, a rigid approach to the use of compensatory measures at early planning stages may prevent projects from moving forward.[89] An administrative body in Flanders recently ruled on the use of compensatory measures in order to avoid a derogation for damage done to nesting sites. In this case, the proponent wanted to establish artificial nests for barn swallows on a newly constructed farmhouse in order to avoid a negative impact on the birds and consequently the need to apply for a derogation. The administrative court (Provincial Authority of Western Flanders) revoked the issued permit as it believed that the creation of artificial nests on the newly rebuilt farmhouse did not prevent the proposed activities from violating the protection rules included in the Species Protection Regulation. Regardless of the ecological soundness of the proposed mitigation measures, the very fact that these were proposed implies that the intention was always to remove the protected nests. As a consequence, a derogation should have been sought (case no. 38014/262/B/2013/111, 4 July 2013).

Although the European Commission clearly excludes the use of compensatory measures to avoid the application of a derogation in the framework of species protection, its position as regard the qualification of mitigation measures is nuanced. According to the Commission, mitigation measures may include actions that actively improve or manage a certain breeding site/resting place so that it does not—at any time—suffer from reduced or lost ecological functionality. These measures can include enlarging the site or creating new habitats in, or in direct functional relation to, a breeding site or resting place as a counterweight to the potential loss of parts or functions of the site for the species.[90] A key requirement here is the continued ecological functionality of breeding sites or resting places for protected species. In view of the limited practice in Flanders, it remains unclear to what extent this

---

[88]See the several cases discussing the adequacy of compensatory measures in wind power plant decisions, Belgian Council of State, 31 July 2014, no 228.147, *Jooris*; Belgian Council of State, 20 September 2011, no 215.210, *Van Laer*; Belgian Council of State, 16 May 2012, no 219.398, *Gatot*; Belgian Council of State, 14 January 2013, no 222.046, *Doudelet*; Belgian Council of State, 11 December 2014, no. 229.530, *Commune de Walhain.*

[89]In the Gatot case mentioned in the previous footnote, the EIA had clearly come to the conclusion that the impact would be significant in the absence of compensatory measures. The conflict may not be apparent in many other instances, as impacts of wind turbines tend to be diffuse and a significant disturbance is difficult to prove. See for example, Belgian Council of State, 14 January 2013, no. 222.046, *Doudelet.*

[90]European Commission, *Guidance Document on the Strict Protection of Animal Species of Community Interest under the Habitats Directive 92/43/EEC*, 47–48.

ecological functionality-approach effectively is implemented whenever issuing planning permits prone to adversely affect strict protected species.

2.2 *The Determination of Biodiversity Offsets and the Use of Metrics*

2.2.1 Ad hoc Metrics and Approaches

In Belgium most biodiversity offsets are currently negotiated and adopted on a case-by-case basis, without relying on specific offset methodologies to measure the losses and gains in a consistent and transparent way. As already discussed, legal provisions rarely set out explicit criteria for the determination of adequate offsets. Factors such as equivalency, additionality, location, timing, and durability are determined on an *ad hoc* basis, reflecting the guidelines in Article 6(4) of the Habitats Directive laid down by the European Commission and/or the scientific literature. Competent authorities generally require like-for-like offsets, meaning that the compensatory measures must target the same species and natural habitat types as those destroyed by the plan or project. However, in most instances such a requirement is not explicitly mentioned in the regulations. In determining the exact scope of offsets, area-based ratios may be set on a case-by-case basis according to the significance of the impact, the potential success of the envisaged compensatory measures, the proximity of the measures, the interim losses and so forth.

The case-by-case determination of biodiversity offsets is subject to judicial control. In cases that find their way to the Council of State, the Court gives particular weight to the opinion of the nature conservation administrations, whether binding or not, in assessing the adequacy of compensation measures.[91] For instance, in the context of wind energy projects in Wallonia, the Council of State recently determined that it is not the role of the judge to reopen technical decisions except in the case of a manifest error of appraisal. The rationale for this limited control is to avoid substituting a legal assessment for the opinion of better qualified expert bodies.[92] However, while it is not the task of administrative courts to second guess the outcome of environmental reports, they are still required to tackle substantive criticism vis-á-vis planning permits and environmental impact studies. In recent years, the Council of State has handed down some seminal rulings as regards the quality requirements for ecological evaluations accompanying planning applications.

In the context of an earlier expansion of the Antwerp Port Area back in 2000, the Council of State rejected the designation of a Natura 2000 site as a compensatory measure because the measure was not additional to the normal implementation of the

[91]See for example, Belgian Council of State, 16 May 2012, no. 219.398, *Gatot*.

[92]Belgian Council of State, 31 July 2014, no. 228.147, *Jooris*.

Habitats and Birds Directives. The Flemish Government had included the designation of an SPA at a location that had already been proposed as a Site of Community Interest as a compensation measure for the construction of the Deurganckdok in the Port of Antwerp. The Belgian Council of State dismissed such approach in its 2002 ruling considering that the Flemish Government had already committed itself to enacting improvement measures at this site by selecting it as a proposed SCI.[93] It could therefore not constitute additional compensatory measures. In a later ruling on the same project development—which was permitted through the application of the derogation clause of Article 6(4) of the Habitats Directive—the Council of State was willing to accept the legality of temporary compensation zones under the guarantee that whenever the zones are abandoned, new compensation zones have been effectively established.[94] Until now, however, the temporary compensation zones are still protected as no additional offset zones have been created in the meantime. Whereas it is a general requirement to ensure the proactive implementation of habitat offsets prior to the encroachment upon nature, the Belgian Council of State seemed inclined to adopt a more lenient approach in one ruling, relating to the extension of the port of Zeebrugge.[95] No explicit attention was paid to likely 'interim losses' in this regard.

2.2.2 Development of Methodologies and Standardized Offset Ratios

The only instances in which methodologies or consistent rules have been established are in the context of deforestation in Flanders and wind energy development in Wallonia.

Forest Offsets in Flanders

As mentioned above, the Flemish Forest Decree lays down a set of specific rules regarding the determination of adequate offsets. In addition, the Forest Protection Regulation stipulates specific rules in regard to the exact implementation of the forest offsets. Interestingly, it establishes standardized offset ratios based on the ecological value of the woodland that will be destroyed through the project development. The decree itself sets minimum ratios starting at 1:1, i.e. an area at least equivalent to the deforested area.[96] This ratio increases to 1:3 if the clearing approval involves a forest that contributes to the conservation objectives of a special area of

---

[93]Belgian Council of State, 30 July 2002, case no. 109.563, *Apers*.

[94]Belgian Council of State, 9 January 2007, no. 166.439, *Apers*; Belgian Council of State, 11 March 2009, no. 191.266, *Apers*.

[95]Belgian Council of State, 16 March 2010, no. 201.909, *Schramme*.

[96]Flemish Forest Decree, art 90bis, § 4, subpara 2.

Table 1 Standardized ratios by square meters and type of forest

Forest type	Ratio
Forest which correspond to one of the following codes: 2160 Dunes with Hippophae rhamnoides 2170 Dunes with Salix repens ssp. argentea (Salicion arenariae) 2180 Wooded dunes of the Atlantic, Continental and Boreal region 9110 Luzulo-Fagetum beech forests 9120 Atlantic acidophilous beech forests with Ilex and sometimes also Taxus in the shrublayer (Quercion robori-petraeae or Ilici-Fagenion) 9130 Asperulo-Fagetum beech forests 9150 Medio-European limestone beech forests of the Cephalanthero-Fagion 9160 Sub-Atlantic and medio-European oak or oak-hornbeam forests of the Carpinion betuli 9190 Old acidophilous oak woods with Quercus robur on sandy plains 91D0 Bog woodland 91E0 Alluvial forests with Alnus glutinosa and Fraxinus excelsior (Alno-Padion, Alnion incanae, Salicion albae) 91F0 Riparian mixed forests of Quercus robur, Ulmus laevis and Ulmus minor, Fraxinus excelsior or Fraxinus angustifolia, along the great rivers (Ulmenion minoris)	3
Other native deciduous forests: comprising at least 80% native deciduous trees	2
Mixed forests: comprising between 20 and 80% native deciduous trees	1.5
Forest of non-native deciduous trees or resinous wood: comprising at least 80% of non-native deciduous trees, resinous trees, or a combination of both	1

Source: Annex 1 of the Forest Regulation

conservation under the Habitats Directive.[97] The Flemish Forest Regulation further specifies the exact ratios that must be applied in relation to the surface area and the tree species composition or according to the presence of a forest habitat type protected at EU level. The regulation includes a table listing the different ratios applicable to each forest type (Table 1). Developers paying into the fund are subject to the same ratios. The total area in square meters that must be reforested/afforested is then multiplied by a certain figure in order to obtain the full price.[98] Currently, this figure is 1.98 EUR per square meter.[99] For more valuable woodlands, this price can rise to almost 4 EUR per square meter. Recently, pleas have been launched to increase the amount of the fee to be paid per square meter of woodland to be destroyed by future development.

The forest compensation, whether in the form of on-the-ground offsets or monetary payments into the fund, must involve the reforestation/afforestation of unforested land (except in rare circumstances) in certain green and public areas

[97]Flemish Forest Decree, art 90bis, § 1, subpara 3.

[98]Flemish Regulation of 16 February 2001 relating to forest compensation and deforestation derogation, Belgian Official Gazette, 23 March 2001 (hereinafter Flemish Forest Regulation), art 5.

[99]Belgian Court of Auditors, *Ontbossing en compensatie. Ultvoering van de compensatieplicht bij ontbossing en werking van het Bossencompensatiefonds, Verslag van het Rekenhof aan het Vlaams Parlement* (Brussels, April 2016), 38.

within applicable zoning plans.[100] In principle, the payment of a monetary forest compensation has to be made within 4 months of the issuance of the permit. The forest compensation *in natura* has to be carried out within 2 years. This can either be realized on land owned by the project developer or that of some third party. If the offsets are not implemented in a timely manner, the permit holder will be liable to criminal sanctions. The non-compliance with permit conditions is to be treated as an environmental criminal offence. Last but not least, the offset site must be maintained as "forest" for a minimum period of 25 years.[101] In other words, there exists no requirement as such to conserve and manage the newly created forests until maturity.

In the last few years, the application of the forest compensation regime has been severely criticised. Recent research has revealed that 'compensation *in natura*' is not always strictly followed and monitored.[102] A positive outcome is only realized for 68% of cases in which 'compensation *in natura*' is applied.[103] There still exists a significant risk that the forest offsets are not effectively implemented, especially in cases of small-scale interventions ('death by a thousand cuts'). In most instances, however, project developers choose financial compensation (60% of cases). Yet, the same research indicates that the financial compensations paid in compliance with the compensation scheme are not adequately used for the afforestation of new parcels of land.[104] To date only 30% of the afforestation target has been achieved (637.2 ha from a total of 2342 ha). The fees levied have not been sufficient to purchase the necessary lands to implement the compensation actions: more than half of the funds generated was used to acquire only 30% of target acreage.[105] The many delays and time-lags that occur when implementing compensation actions partly explain the limited success until now of the forest compensation regime in halting the net loss of forest habitats in the Flemish Region in terms of quantity.[106] Let alone that the protection schemes are adequate in ensuring that the offsets achieve the former ecological quality of the lost woodlands.

Methodological Framework for the Establishment of Wind Farms in Wallonia

In 2012 the administration of the Walloon Region developed a methodological framework for the determination of appropriate mitigation measures and biodiversity

[100] Flemish Forest Regulation, art 4, § 3.
[101] Flemish Forest Regulation, art 4, § 3.
[102] Belgian Court of Auditors, 2016.
[103] Ibid, 45.
[104] Ibid, 57.
[105] Ibid, 58.
[106] See also: "Heeft Schauvliege dan nog geen dossiers ingediend voor boscompensatiefonds?", *De Standaard* 21 Oktober 2015.

offsets in the context of wind power development projects.[107] Given the important number of wind energy projects in the region and the likely conflicts with bats and birds species, the sector was in need of guidelines on how to integrate biodiversity issues in their projects. The traditional case-by-case assessment was proving to be a considerable drag on the approval process, forcing developers to amend their applications on multiple occasions to satisfy the administrative requirements. In addition, the various field offices of the nature conservation agency often issued conflicting opinions on the acceptability of impacts and the adequacy of proposed measures. Rather than setting strict rules, the methodological framework is an internal document designed to guide developers and ecological consultants in proposing biodiversity offsets, and guide field offices in assessing the project proposals. Its aim is to bring consistency, coherence, objectivity and efficiency to the process, thereby avoiding a strict case-by-case approach.

The methodological framework focuses on bats and birds most likely to be impacted by wind turbines. It establishes a checklist of recommended steps to assess the ecological value of the site, from identifying which natural habitats and species are present to proposing various mitigation and compensation measures. The methodology sets out the preferred data collection and field survey protocols for each species group likely to be impacted (wintering, migrating and nesting birds as well as bat species). The goal is to evaluate the potential of the site as a habitat for each bird and bat species. Subsequently, the sites are categorized into three risk (concern) levels for bats and four levels for birds. For bats, the level of risk is determined by the site and is based on the threatened status of species found there as well as on the proximity of forests. For instance, if the site is inhabited by a Natura 2000 species, any disturbance will be classified as 'major risk' for that species. For birds, the level of risk is determined for each key species present on a particular site and is based on the local abundance of the species and the location in regard to the regional range of the species. The level of risk determines whether the impact can be considered acceptable as well as which measures should be envisaged (mitigation and offsets) (Table 2).[108]

The framework includes a list of standard compensatory measures to be applied according to the group of species at stake and the level of risk (Table 3). These are based around two umbrella species (harriers and grey partridge). It is considered that other species will benefit from the same type of management activities implemented for these umbrella species. The types of measures are: maintenance of feeding vegetation cover in winter (COA1); creation of permanent grassy headlands

[107]Jérémy Simar and others, *Projets éoliens: Note de référence pour la prise en compte de la biodiversité* (DGARNE, Septembre 2012, 2012). The framework was developed by the Natural and Agricultural Environmental Studies Department and the Wildlife and Forestry Department within the Operational Directorate-General for Agriculture, Natural Resources and the Environment (DGARNE). It can be consulted at http://biodiversite.wallonie.be/servlet/Repository/28103.pdf?ID=28103

[108]Note that the framework considers offsets unwarranted for bat species protected at the European level as there is currently little reliable expertise in the restoration of bat colonies.

Table 2 Level of risk and measures recommended for non-breeding birds

Non-breeding birds: wintering			
	Level of risk	Mitigation measures	Compensation measures
Occurrence of the Northern Harrier feeding on the site			
1 occasional individual	Medium	Do not facilitate or reduce public access	Not required
1 regular individual	Strong	Do not facilitate or reduce public access	Not required
Several occasional individuals	Strong	Do not facilitate or reduce public access	COA1 & COA2
Several regular individuals	Major	Do not facilitate or reduce public access	COA1 & COA2
Occurrence of the Marsh Harrier feeding on the site			
1 occasional individual	Medium	Do not facilitate or reduce public access	Not required
1 regular individual	Strong	Do not facilitate or reduce public access	Not required
Several occasional individuals	Strong	Do not facilitate or reduce public access	COA1 & COA2
Several regular individuals	Major	Do not facilitate or reduce public access	COA1 & COA2
Occurrence of Red Kite feeding on the site or within 1000 m of wind turbines			
1 occasional individual	Low	Do not facilitate or reduce public access	Not required
1 regular individual	Medium	Do not facilitate or reduce public access	Not required
Several occasional individuals	Medium	Do not facilitate or reduce public access	Not required
Several regular individuals	Strong to major	Do not facilitate or reduce public access	Currently not offsetable

Source: Simar J and others, Projets éoliens: Note de référence pour la prise en compte de la biodiversité (DGARNE, Septembre 2012, 2012), 66 (extract).

(COA2); herbaceous buffers along access roads and crane hard standings (COA3); buffer zones along streams (COA4); hedgerows featuring a grassy bench (COA5); measures to encourage the nesting of Northern Lapwings (COA6). The amount of offsets is then determined from the specific diversity of the site and the level of risk for each species. Instead of applying a particular ratio to the whole site impacted by the wind farm, the framework establishes a certain number of hectares on which to implement the compensatory measures by disputed wind turbine (Table 4). At the time of writing, it is too early to assess the effectiveness of the methodology in achieving adequate offsets. One apparent flaw is the fact that the offset site is still chosen on a case-by-case basis, in response to opportunities in terms of land access rather than to maximize biodiversity performance.

Table 3 Adequate measures for each species

	COA1	COA2	COA3	COA4	COA5	COA6
Northern Harrier (Circus cyaneus)	X	X				
Marsh Harrier (Circus aeruginosus)	X	X				
Montague's Harrier (Circus pygargus)	X	X				
Corn Bunting (Emberiza calandra)	X	X*	X*	X*	X*	
Yellow-hammer (Emberiza citronella)	X	X*	X*	X*	X*	
Western Yellow Wagtail (Motacilla flava)	X*	X*	X*	X*	X*	
Northern Lapwing (Vanellus vanellus)						X
Skylark (Alauda arvensis)	X	X*				
Short-eared Owl (Asio flammeus)	X	X*				
Common Quail (Coturnix coturnix)	X*	X*	X*	X*	X*	
Grey Partridge (Perdix perdix)	X	X	X	X	X	

X*: species not directly targeted by the measure but which indirectly benefit from it
Source: Simar J and others, Projets éoliens: Note de référence pour la prise en compte de la biodiversité (DGARNE, Septembre 2012, 2012), 89.

Table 4 Ratios for breeding birds

• If at least one of the studied risks is major, the area targeted by the compensatory measures is 3 ha for each disputed wind turbine, with the possibility to advise against the establishment of the disputed turbines
• If none of the studied risks is major but at least one is strong, the ratio is 2 ha for each disputed turbine. If a majority of the studied risks are strong, the option to advise against the establishment of the disputed turbines should be considered
• All other configurations of risk lead to a ratio of 0–1 ha for each wind turbine

Source: Simar J and others, Projets éoliens: Note de référence pour la prise en compte de la biodiversité (DGARNE, Septembre 2012, 2012), 106–107.

2.3 *Effectiveness of Biodiversity Offsets*

Article 64, § 4, of the Nature Conservation Ordinance empowers the government in Brussels to take all necessary actions to ensure the effectiveness of biodiversity offsets. For example, the government may require that the compensatory measures be effective before the commencement of the work, order the applicant to monitor the implementation of the measures, designate the compensatory areas as Natura 2000 sites, and expropriate land ecologically suited to implementing the required measures. In addition to these actions aimed at ensuring the effectiveness of individual compensatory mitigation projects, the government has a general obligation to monitor the effectiveness of the measures at the regional level. In particular, the Ordinance establishes a comprehensive conservation planning framework in which the government must monitor biodiversity, publish regular reports on the state of biodiversity, establish a regional nature conservation strategy, and periodically review the effectiveness of the conservation actions undertaken.[109] Within the review, the government is obliged to monitor the compensatory measures adopted in derogatory regimes established in the Ordinance.[110] The result of the review should inform the adoption of the nature conservation strategies.

Contrary to the Brussels Nature Conservation Ordinance, neither the Flemish Nature Conservation Decree nor the Walloon Nature Conservation Act contains any specificities with respect to offset monitoring and enforcement. This situation may be revised in the coming years, especially in the wake of stricter monitoring requirements introduced to the EIA Directive by its 2014 review. Moreover, by pushing for including as much information as possible in the conditions of permits, the case law of the Belgian Council of State seems to underscore the important duty of the government when it comes to the monitoring and enforcement of future offsets. The Belgian Council of State has consistently held that conditions included in the permit must be strictly complied with and not obeyed at the discretion of the permit holder.[111] A competent authority cannot approve a project on the condition that a compensatory plan will be developed in the future, or if the implementation of the measures is uncertain or depends on a third party. This means that the details of compensatory measures are to be fixed beforehand,[112] and must include specification of the exact type of measures, their physical extent in hectares, their location and their duration.[113] This can then be monitored. In 2013 the Council of State revoked a permit in part because it did not include information as regards the duration and monitoring of compensatory measures.[114]

Moreover, if the offsets are to be implemented by a third party, the Council of State requires that contracts between the permit seeker and the party implementing

---

[109]Dupont and Born, 88.

[110]Brussels Nature Conservation Ordinance, art 15, § 2, 6°.

[111]Belgian Council of State, 20 September 2011, no. 215.210, *Van Laer*; Belgian Council of State, 16 May 2012, no. 219.398, *Gatot*; Belgian Council of State, 14 January 2013, no. 222.046, Doudelet. The same reasoning applies to environmental permits, see Belgian Council of State, 11 December 2014, no. 229.530, *Commune de Walhain* (environmental permits) On this question, see Monique Kestemont, Michel Karounski and Frédéric De Munck, 'Les Conditions Assortissant les Permis – Etat des Lieux' (2011) Aménagement, Environnement, Urbanisme et Droit Foncier 249.

[112]Belgian Council of State, 20 September 2011, no. 215.210, *Van Laer*. In this case, the competent authority had approved a wind power plant while requesting that compensatory measures be implemented to strengthen the connectivity between three identified natural entities, including a Natura 2000 site (Vallée de la Burdinale), but leaving the exact determination of the measures to a scientific committee to be set up after the approval. The Council of State ultimately annulled the permit because the competent authority did not have all the elements necessary to make a decision and conditioned the execution of the permits on the intervention of a third party.

[113]Belgian Council of State, 11 December 2014, no. 229.530, *Commune de Walhain*. On the exact number of hectares, see also Belgian Council of State, 31 July 2014, no. 228.147, *Joris*. See in a similar vein: Belgian Council of State, no. 224.750, Vanmassenhove; Belgian Council of State, no. 227.223.

[114]Belgian Council of State, 14 January 2013, no. 222.046, *Doudelet*.

the measures be signed before the permit approval and referred to in the permit.[115] Similarly, in the framework of forest offsets in Flanders, the project developer wishing to implement an obligation through a third party will have to demonstrate that s/he has obtained the necessary guarantees in this respect and document them in the permit application. In Belgium, a country characterized by a high level of urban development, finding appropriate compensation areas is often a very laborious task. Therefore, offsets should be backed by proof of ownership or a contract with a willing landowner. More recently, the Council of State somewhat loosened its case law by asserting a newly developed administrative practice, according to which the proof of ownership is no longer required in advance. This was deemed legal since the execution of the permit had been made conditional on the effective realization of the offsets.[116] In other words, if the project developer is unable to secure suitable lands to implement the requested compensation, the project will not be able to move ahead. This new decision indicates that the Council of State focus on the means to ensure the effectiveness of the measures, whether it be advanced proof of land access or conditional execution of the permit. As indicated, the increasing focus on the successful implementation of offsets is understandable given the poor track record to date in this respect.

From a legal perspective, the enforceability of the measures is crucial to ensure their effectiveness. Thus, the offsets measures should be clear-cut and precise as to their content and objectives, and the responsible parties for their implementation must be identified. This will allow efficient scrutiny as well as the application of penalties if permit conditions are breached. In most planning permits, the compensatory measures are expressed in terms of actions to undertake rather than performance standards to be achieved. In other words, the condition of a permit is met once the action is implemented in accordance with the offset plan. In some instances, this will include a one-off measure such as the reforestation of a site. In others, the compensatory actions will have to be conducted periodically over a period of time.

Ideally, the offset package should include a monitoring programme. In this respect, following the 2014 revision of the EIA Directive (yet to be transposed into regional laws), it is now the case that environmental conditions included in a permit as a result of the environmental impact assessment must contain monitoring measures.[117] For projects of a certain size, a monitoring scientific committee may be established to oversee the implementation of actions.[118] Yet, when permits do not

---

[115]Belgian Council of State, 11 December 2014, no. 229.530, *Commune de Walhain.* Note that, except in the case of financial compensation for impacts on forests, the proponent of the project is liable for the implementation of measures even if the measures are implemented by a third party. To date, Belgian law does not have a mechanism in which the offset operator could be held liable for failing to implement the compensatory measures contained in a permit. They are, however, bound by contract law.

[116]Belgian Council of State, 7 October 2015, no. 232.475, *Bollinger.*

[117]EIA Directive, art 8bis(1)(b).

[118]First Book of Walloon Environmental Code, art D 29-25. See C.E., 20 September 2011, Van Laer, n° 215.210. See also Born, Dupont and Poncelet, 34.

expressly identify performance standards as discussed above, it may be hard to set up an efficient monitoring programme that can provide information on possible adaptations. Furthermore, the law is not suitably equipped for adaptive management. The only way to modify the compensatory measures would be to amend the permit.[119] Interestingly, when adopting a revised zoning plan for the further extension of the port of Antwerp that would entail the partial destruction of several Natura 2000 sites, the Flemish Government made the execution of the planning permit contingent on the effective implementation of the biodiversity offsets on the ground. Only if the competent Flemish authorities deem the creation of the new nature ‘core areas’ to be successful can the construction works go ahead. However, given the fact that the latter measures had been qualified as ‘mitigation’ rather than ‘compensation’, the viability of this approach remains subject to the outcome of the pending lawsuits (see supra).

3 Practice in Belgium

3.1 Case Study in Belgium: Biodiversity Offsets in Flanders

Several case studies have been presented in the preceding analysis. Most noteworthy are the judicial decisions regarding the use of mitigation and compensation measures in the context of port development with a negative impact on Natura 2000 sites. However, in 2015 a major controversy arose around the expansion of the premises of a transport company (*H.* Essers) in a Natura 2000 site. To some extent, this case epitomizes many of the shortcomings of the Flemish biodiversity offsetting policies over the past decades. The case provoked a number of editorials in the national press.[120] For one well-known Flemish comedian this was a blatant illustration of the lack of political interest in preserving Flanders’ most valuable biodiversity.[121] It was the first time that the issue of biodiversity offsets hit the headlines of the national press. The discussion primarily focused on the mitigation and compensatory duties that have to be observed when a project development will impact on a Natura 2000 site. In order to understand the relevance of the case, some background information is required. The transport company H. Essers has its headquarters and primary activities located in the municipality of Genk, in the Flemish province of Limburg. In 2007 the company wanted to expand its activities through the construction of a new transport hall. Since the industrial estate bordered a Natura 2000 site, the

[119]See for example Walloon Environmental Permit Decree, art 64ff.

[120]See for example: Hendrik Schoukens, ‘There’s something rotten in Flanders’ forests...’, *De Morgen*, 27 September 2015; Hendrik Schoukens, ‘Wie doet Essers wat?’, *De Standaard*, 19 January 2016.

[121]See for instance: http://www.demorgen.be/binnenland/wouter-deprez-bijzonder-pijnlijk-dat-schauvliege-het-sleutelargument-in-het-dossier-niet-kent-bd952f95/

expansion would inevitably lead to the immediate destruction of 1.7 ha of protected woodland habitat. The appropriate assessment carried out prior to the planning permit suggested that, in order to offset the project's significant impact, 10 ha of degraded woodland would have to be restored. The restoration zone bordered the existing industrial estate and was located within the boundaries of the Natura 2000 site. The restoration measures included the felling of conifers and the planting of thousands of young birches and oaks. No application of the derogation clause had taken place. The restoration measures were wrongly tagged as 'mitigation measures' while they ought to be qualified as compensation. The so-called mitigation plan also foresaw the re-establishment of heather. In addition, the planning permit concluded that the company was to be granted no further expansion in coming years in order to preserve the integrity of the Natura 2000 site. The necessary planning permits were issued in 2009 and translated into a general zoning plan. Although some environmental NGOs voiced their concerns over the adverse effects of the project, no legal challenge was launched before national courts aimed at contesting the non-application of the derogation clause in this case.

However, notwithstanding the irregularities surrounding the exact legal qualification of the so-called 'offset zone', it was also poorly implemented on the ground. No trees were planted in the so-called mitigation zone nor was the heather re-established. The zone itself was in the hands of the local municipality, which failed to revise its forest management plan in order to facilitate the requested mitigation. At no point over the years have the competent authorities begun criminal proceedings against the permit holder for not complying with the mitigation requirements explicitly included in the permit conditions. However, in 2014 the company initiated yet another planning procedure in order to further expand its transport activities. This time the company wanted to develop the mitigation zone it had previously failed to restore. Remarkably, the competent authorities decided to approve the new planning permit, ignoring the non-compliance of the earlier restoration duties on the part of the company. The appropriate assessment failed to focus on the cumulative impact of the successive impairments. By contrast, it rather succinctly concluded that no significant effects were to be expected given the fact that the expansion site did not contain actual patches of well-developed habitat. Habitat restoration measures were stipulated for several of the protected species whose habitat would be partially destroyed as a result of the development actions. Also, it was argued that the project would ultimately lead to a win-win situation for nature given the forest offsets that had been provided elsewhere in the province of Limburg. These offsets, however, were not specifically tailored to mitigate or compensate the immediate loss of Natura 2000 habitat resulting from the new expansion. Several environmental NGOs spoke out against the government's final approval of the development, initiating several lawsuits to challenge the legality of the zoning plan, which, finally adopted in April 2016, gave the green light for the further economic development of the site.

The contents of the permits and zoning plans lay bare many of the shortcomings of the Flemish mitigation policy. Specifically, these are as follows:

Continued Net Loss The case study clearly reveals the blatant lack of enforcement and monitoring of existing mitigation commitments, even in the context of strictly protected Natura 2000 sites. In fact, on several occasions the competent minister explicitly acknowledged that the restoration measures provided for in the 2009 planning permit had not been complied with. Most surprisingly, this non-compliance was by no means an obstacle to the further development of the site. While the refusal to observe mitigation measures should give rise to criminal prosecutions, the present case study clearly shows that this is more a theoretical than practical consideration. In view of the limited enforcement of mitigation duties, even in the context of strictly protected biodiversity, it is easy to understand why the mitigation schemes fail to deliver, and why we observe a continued net loss of biodiversity.

Inadequate Application of the Mitigation Hierarchy Frequently, habitat restoration measures are simply labeled 'mitigation' in order to justify the issuance of a permit in the context of a vulnerable Natura 2000 site without having to take recourse to the more stringent derogation clause. In particular, the so-called mitigation measures included in the 2009 planning permit, namely the planting of trees and re-establishment of heather, clearly qualified as compensation. Such measures should only be considered when less harmful project alternatives are unavailable and the project itself qualifies as an 'imperative reason of overriding public interest'. In the current case this balancing of interest never took place, indicating that the mitigation hierarchy is often disregarded, even in the context of EU protected sites. It appears that 'paper' restoration measures are often used as a cover-up for unsustainable project development. The many uncertainties surrounding ecological restoration efforts should make authorities hesitant to accept these as forms of mitigation. In practice, however, habitat restoration is now used as 'panacea for all ills'.

Collateral and Cumulative Loss As demonstrated, no specific attention has been paid to the cumulative impact of the successive economic developments. Given the explicit rules on the avoidance of cumulative effects in the context of Natura 2000, these findings are rather worrying. It can be expected that collateral and cumulative losses are generally ignored in the context of project developments, especially in the context of losses to biodiversity not enjoying strict protection under Flemish nature conservation law. In most instances, such cumulative effects remain outside the scope of ecological assessments.

Reference Loss In 2016 the second expansion of the transport company was justified by pointing out the limited ecological potential of the area of the Natura 2000 site affected by the new project development. However, the applicable conservation objectives indicated that the Natura 2000 site was currently in an unfavorable conservation state partly due to the uncontrolled urban development in the immediate surroundings. In many instances the extent of mitigation measures or offsets is kept artificially low by highlighting the already degraded status of the affected biodiversity.

Baseline Loss The mitigation measures included in the 2009 planning permits did not fully offset the biodiversity loss related to the projected developments. Indeed, instead of aiming to restore a site with no actual biodiversity assets, the mitigation zone already contained valuable biodiversity. The mitigation measures should be directed to converting the site into woodland of greater ecological value. Furthermore, it can be argued that such measures should exceed the existing commitments for the zone. Given its location in a Natura 2000 site, the competent authorities were already under an obligation to restore it to a favourable conservation status.

In its provisional ruling of July 4th 2017, the Belgian Council of State decided to suspend the spatial execution plan because the appropriate assessment failed to take into account the cumulative effects caused by the first expansion and the non-implementation of the biodiversity offsets linked thereto.[122] In addition, it ruled that the restoration measures aimed at offsetting the loss of habitat of an Annex II-species under the Habitats Directive could not be qualified as 'mitigation' and thus application should have been made of the derogation clause. Through the prism of the obligation to adequately assess cumulative effects, the Council was able to sanction the failure to properly compensate the previous biodiversity loss caused by the first expansion of the company. Indirectly the Belgian Council of State thus took into consideration the faltering implementation of the previous biodiversity offsets as an argument to invigorate the duty to adequately assess the overall impact of the novel expansion plans. If reaffirmed in subsequent rulings, this case might function as a major turning point for the proper enforcement of biodiversity offsets in a spatial planning context.

4 Conclusions from Belgian Practitioner's Perspective on an EU-Wide No Net Loss and Offset Strategy

The recent popularity of biodiversity offsets as a policy instrument to halt the increased loss of biodiversity caused by economic development has attracted heated debate. For some government agencies and industries, offsets are heralded as a means to reconcile development with nature conservation. Proponents see many benefits: biodiversity offsets respect the "no net loss" principle without hampering further sustainable development, they serve to increase financial resources for the conservation of biodiversity, they give a sense of responsibility to economic actors, and can help to manage environmental risks. Critics, however, who include many environmental NGOs, argue that the wider use of biodiversity offsets might lead to the commodification of nature, sending out the dangerous message that natural

[122] Belgian Council of State, 4 July 2017, no. 238.763, *vzw Bond Beter Leefmilieu Vlaanderen.*

resources can be replaced if only enough money is available ("licence to trash").[123] Between these opposing views, the scientific community has adopted a cautious position, largely identifying challenges and trying to offer solutions. In recent literature, clear warnings have been expressed concerning the often too optimistic premises upon which such offsetting schemes are grounded.[124]

Until recently, no widespread debate had been conducted in Belgium on the pros and cons of biodiversity offsets. This can perhaps be attributed to the lack of strict enforcement. Even if offsets were specified in planning permits, no strict monitoring of the restoration commitments subsequently took place. The recent rise in the number of lawsuits centering on nature protection has corrected the lax attitude towards biodiversity offsetting, which previously was often viewed as an easy solution for harmful project development. This is reflected in more rigorous judicial scrutiny, whereby judges are frequently dismissing lenient approaches to compensation. Yet while biodiversity offset schemes should certainly not be presented as "panacea to all ills", they do have the potential to promote sustainable development and to provide developers and decision-makers with an incentive to better consider the biodiversity impacts of their development projects. If applied within a strict regulatory framework, biodiversity offsets can indeed serve as an instrument to balance conflicting interests—for instance economic development and nature protection—in a transparent and accountable way. By requiring developers to compensate their impacts, competent authorities force them to take into account the cost of restoration or replacement measures, thereby internalizing the negative impacts which their projects have on biodiversity.[125] In that sense, biodiversity offsets may play a very important preventive role, even more important than the effective remediation to authorized environmental damages.

Interestingly, as early as 2006 the Belgian National Biodiversity Strategy (adopted at the federal level in collaboration with regional authorities with a view to meeting the pledges made under the Convention on Biological Diversity) established compensation as a guiding principle of environmental actions alongside the precautionary principle, the principle of preventive action and the polluter pays principle. In addition to strategic and operational objectives, the Strategy lays down 10 fundamental principles of environmental law that should guide the environmental action of the regions and the Federal Authority.[126] In particular, Principle 10 of the

[123]http://naturenotforsale.org/letter2eu/

[124]See among others: David Moreno-Mateos, Virginie Maris, Arnaud Béchet and Michael Curran, 'The true loss caused by biodiversity offsets' (2015) Biological Conservation 552; Martine Maron and others, 'Conservation: stop misuse of biodiversity offsets' (2015) Nature 401.

[125]Born, Dupont and Poncelet 12.

[126]Belgium's National Biodiversity Strategy 2006-2016; Belgian National Focal Point to the Convention on Biological Diversity (ed.), 2013. Biodiversity 2020 – Update of Belgium's National Biodiversity Strategy. Royal Belgian Institute of Natural Sciences, Brussels, 148 pp (hereinafter 'updated national biodiversity strategy').

Strategy states that if a plan or project must proceed for imperative reasons of overriding public interest despite a negative assessment of the implications for biodiversity and in the absence of alternative solutions, then public authorities should take all compensatory measures necessary to ensure that no net loss of biodiversity will occur when the plan or project is implemented or executed. Following a mid-term assessment of the current situation, the strategy was amended in 2013. An Operational Objective 3.8 was added to define "the framework and the conditions to ensure no net loss of biodiversity and ecosystem services".

By establishing biodiversity compensation as a guiding principle, the federal strategy should push regional authorities to further implement compensatory measures in their decision-making process regarding projects with significant impacts on biodiversity, including unavoidable residual impacts on species, habitats and ecosystem services not covered by Natura 2000. However, given the non-binding nature of the strategy, it remains to be seen whether this can really serve as a trigger to foster debate on biodiversity offsetting at the regional level. At the programmatic stage, the Brussels nature conservation plan adopted on April 14th 2016 aim to develop tools and procedures to better integrate biodiversity concerns in plans and projects. In order to allow regional development to continue without generating a net loss of biodiversity, a study will be launched on the desirability and the implementation modalities of a compensation mechanism in situ across the region, with landscape or ecosystem units to be defined.[127] In the Walloon Region, the government recently developed a catalogue of actions to be undertaken by stakeholders.[128] This also places more focus on integrating biodiversity loss in the decision-making process. If necessary, offsets may be permitted to avoid further biodiversity loss. In Flanders there has been no wider debate on the desirability of biodiversity offsetting. Only recently has attention shifted to the relatively poor achievement of the no net loss-policy in the field of forest protection (see the case study above). Also the application of biodiversity offsetting in the context of Natura 2000 has recently been discussed. In spite of the Strategy's identification of offsets as an important policy instrument, no clear-cut policy objectives have been implemented as to the use of biodiversity offsetting as a means of halting biodiversity loss.

To date, the practice of biodiversity offsets in Belgium has been mainly driven by EU law. Indeed, the most telling cases of biodiversity offsetting relate to projects that affect EU-protected sites. The limited offsetting schemes that apply to biodiversity *not* protected under EU law are often poorly enforced. One worrying trend is the use of biodiversity offsets as a tool to bypass the strict derogation procedures imposed by EU law, such as the Habitats Directive, for unsustainable development project. Obviously, such a tendency risks further undermining the premises upon which

[127] Brussels Government, *Brussels Regional Nature Plan 2016–2020* (14 April 2016), Action 9.3.

[128] Walloon Government, *Network Wallonia Nature, Catalogue of actions, version III* (February 2015).

biodiversity offsetting is founded. Furthermore, in the absence of comprehensive guidelines, biodiversity offsets are mostly developed on a case-by-case basis with little coherence, efficiency and consistency. Belgium has moreover one of the highest population densities in the world, which increases the cost of achieving no net loss as project proponents often have trouble finding suitable land to implement offsets and maintain them in the long run.

Logically, the development of a coherent no net loss policy at the European level could further streamline the practice of biodiversity offsetting in Belgium by serving as a template for the refinement of offsetting schemes at the regional level. Ideally, the core elements of any offsetting strategy—equivalency, additionality and longevity—should be established as guiding principles of biodiversity offsetting strategies at regional level. Only if these requirements are observed will biodiversity offsetting schemes be capable of delivering the expected benefits or, at the very least, avoid a further loss of biodiversity. In this regard the Flemish practices demonstrate that only a strictly regulated biodiversity offset regime administered by a specialized agency is capable of guaranteeing no further loss of biodiversity. A reliable methodology for determining credit exchange is currently lacking in Belgium. In this regard, the further development of standard metrics as done for forest clearing and wind development impacts would greatly enhance the quality of biodiversity offsets in Belgium. This represents but one of the many challenges that need to be faced.[129]

References

Alen A et al (1992) Treatise on Belgian constitutional law. Kluwer Law and Taxation, Boston

Anker HT (2014) Simplifying environmental legislation – reviewing the eia directive? J Eur Environ Plann Law 11(4):321–347. https://doi.org/10.1163/18760104-01104002

Belgian Court of Auditors (2016) Ontbossing en compensatie. UItvoering van de compensatieplicht bij ontbossing en werking van het Bossencompensatiefonds. Verslag van het Rekenhof aan het Vlaams Parlement, Brussels

Bocken H, De Bondt W (2001) Introduction to Belgian law. Kluwer Law International, The Hague

Born CH (2015) Le diable dans les détails: les défis de la régulation des marchés d'unités de biodiversité. Revue internationale de droit économique 2:151–182. https://doi.org/10.3917/ride.292.0151

Born CH, Dupont V, Poncelet C (2012) La Compensation Ecologique des Dommages Causés à la Biodiversité: un Mal Nécessaire? Aménagement, Environnement, Urbanisme et Droit Foncier 3:12–40

Brussels Government (2016) Brussels Regional Nature Plan 2016–2020

De Smedt P, Schoukens H (2009) Natuurzorgplicht: is er leven na het arrest van het Grondwettelijk Hof van 27 mei 2008? Niew Juridisch Weekblad 210:738–758

Dupont V, Born CH (2014) L'ordonnance du 1er mars 2012 relative à la conservation de la nature: une avancée significative pour la préservation de la biodiversité urbaine et périurbaine à

[129]Charles-Hubert Born, 'Le diable dans les détails: les défis de la régulation des marchés d'unités de biodiversité' (2015) Revue internationale de droit économique 151.

Bruxelles (Partie II). Aménagement, Environnement, Urbanisme et Droit Foncier: revue d'études juridiques 3:138–152
Dupont V, Schoukens H, Cliquet A (2014) Aspects juridiques de la restauration écologique: un avenir du droit de la conservation de la nature? Aménagement, Environnement, Urbanisme et Droit Foncier 4:84–102
EFTEC, IEEP (2010) The use of market-based instruments for biodiversity protection-the case of habitat banking, Technical Report for European Commission DG Environement, London
European Commission (2007a) Guidance document on Article 6(4) of the 'Habitats Directive' 92/43/EEC
European Commission (2007b) Guidance document on the strict protection of animal species of community interest under the Habitats Directive 92/43/EEC
European Commission (2011) Communication from the Commission to the European Parliament, the Council, the Economic and Social Committee and the Committee of the Regions. Our life insurance, our natural capital: an EU biodiversity strategy to 2020 (COM(2011) 244 final)
European Environment Agency (EEA) (2015) The European environment – state and outlook 2015
Kestemont M, Karounski M, De Munck F (2011) Les Conditions Assortissant les Permis – Etat des Lieux. Aménagement, Environnement, Urbanisme et Droit Foncier 249–164
Leakey RE, Lewin R (1995) The sixth extinction – patterns of life and the future of Humankind. Doubleday, New York
Maron M et al (2015) Conservation: stop misuse of biodiversity offsets. Nature 523:401–403. https://doi.org/10.1038/523401a
Millenium Ecosystem Assessment (2005) Ecosystems and human well-being: biodiversity synthesis
Moreno-Mateos D, Marris V, Béchet A, Curran M (2015) The true loss caused by biodiversity offsets. Biol Conserv 192:552–559
Schoukens H (2014a) Natuurbescherming Buiten de Lijntjes: de Algemene Natuurtoets als Imperfect Antwoord op het Verlies aan Natuur binnen Vlaanderen. Tijdschrift voor Omgevingsrecht en Omgevingsbeleid 3:346–370
Schoukens H (2014b) Région Flamande. In: Born C-H, Haumont F (eds) Natura 2000 et le juge – Situation en Belgique et dans l'Union européenne/Natura 2000 and the Judge – Situation in Belgium and in the European Union, 1rd edn. Bruylant (actes du colloque organisé à Louvain-la-Neuve, le 31 mars 2011)
Schoukens H (2014c) De ondraaglijke lichtheid van milieueffectrapportage: Einmal ist Keinmal? Tijdschrift voor Ruimtelijke Ordening en Stedenbouw 73:20–41
Schoukens H (2017) Habitat restoration measures as facilitator for economic development within the context of the EU habitats directive: balancing no net loss with the prevention approach? J Environ Law 29(1):47–73
Schoukens H, Cliquet A (2014) Mitigation and compensation under EU nature conservation law in the Flemish region: beyond the deadlock for development projects? Utrecht Law Rev 10:194
Schoukens H, Desmedt P, Cliquet A (2007) The implementation of the habitats directive in Belgium (Flanders): back to the origin of species? J Eur Environ Plann Law 4(2):127–138
Schoukens H, De Roo K, De Smedt P (2011) Handboek Natuurbehoudsrecht, vol 64. Kluwer, Mechelen
Simar J et al (2012) Projets éoliens: Note de référence pour la prise en compte de la biodiversité. Septembre, DGARNE, available at http://biodiversite.wallonie.be/servlet/Repository/28103.pdf?ID=28103 (last accessed on 30 December 2017)
Steichen P (2009) Le principe de compensation, un nouveau principe du droit de l'environnement? In: La responsabilité environnementale, prévention, imputation, réparation
Van Hoorick G (2014a) Compensatory measures in European nature conservation law. Utrecht Law Rev 10:161
Van Hoorick G (2014b) Evolution Récente du Droit de la Conservation de la Nature en Région Flamande. Aménagement, Environnement, Urbanisme et Droit Foncier 32
Walloon Government (2013) Reference framework for the installation of wind turbines in Wallonia (approved on 21 Feb 2013)
Walloon Government (2015) Network Wallonia nature, catalogue of actions, version III (Feb 2015)

The Czech Republic

Adéla Boučková and Juliane Albrecht

1 Introduction

The Czech Republic is a landlocked state in central Europe with a population of approx. 10.5 million. It has a centralised administrative structure. Almost all borders to the four neighbouring countries (Germany to the west, Poland to the north, Slovakia to the east and Austria to the south) are marked by low mountain ranges. In terms of physical geography, the Czech Republic features landscapes of rolling hills and river basins ringed by mountains.

Despite its relatively small landmass of 78,866 km^2, the country has a diverse physical geography with numerous valuable plant and animal species. Czech law provides for various instruments to protect nature and biodiversity. Some of these are based on European law, introduced at the time of the Czech Republic's accession to the European Union in 2004.

However, Czech environmental law does not provide for any instrument of compensation covering the entire country, e.g. one corresponding to Germany's *Eingriffsregelung* (Impact Mitigation regulation). At the same time, duties of compensation are determined for individual sectors such as within the Law on Nature and Landscape Conservation (ZOPK 1992) as well as other laws.

A. Boučková (✉)
Charles University, Prague, Czech Republic
e-mail: a.bouckova@gmail.com

J. Albrecht
Leibniz Institute of Ecological Urban and Regional Development, Dresden, Germany
e-mail: j.albrecht@ioer.de

W. Wende et al. (eds.), *Biodiversity Offsets*,
https://doi.org/10.1007/978-3-319-72581-9_6

2 Compensatory Instruments in the Law on Nature and Landscape Conservation

Obligations to provide compensation for interventions in nature and landscapes are, in the first instance, regulated by the Law on Nature and Landscape Conservation (zákon č. 114/1992 Sb., o ochraně přírody a krajiny, "ZOPK").

2.1 Protection of Trees Outside Woodland (§§ 7–9 ZOPK)

The most important instrument of compensation aims to protect trees and bushes outside woodland areas. This is regulated by §§ 7–9 ZOPK. Under Czech nature protection law, the conservation of trees and bushes outside woodland is governed by general species protection, under which trees and bushes are protected against damage or destruction (General Prohibition of Deterioration, § 7 para. 1 ZOPK).

Above a certain size, official permission must be sought for tree felling (§ 8 para. 1 ZOPK). The Decree No. 189/2013 of the Ministry of the Environment designates those trees and bushes which are to be protected. Trees that may be felled without seeking prior permission are those with a trunk circumference up to 80 cm (measured at a height of 130 cm above ground) as well as plantation trees and fruit trees in gardens, insofar as these are not parts of protected landscape elements as determined by § 3 para. 1 b) ZOPK.

Permission for felling can only be granted after an evaluation of the functional and aesthetic significance of the trees and in response to some compelling reason. Compelling reasons are limited to cases where the intended aim of the tree felling (e.g. to construct a building or a road) cannot be achieved by any other means (Stejskal 2016, p. 95).

If permission for felling is granted, the responsible environmental protection agency can order **suitable compensatory plantings** (§ 9 para. 1 ZOPK) to offset the environmental damage. Furthermore, the agency can order that the compensatory plantings be tended for a requisite period of up to 5 years.

Regarding the form and extent of compensatory plantings, administrative organs have a certain margin of discretion. Hence, the nature conservation authorities can determine which species and number of trees must be planted. If construction work will be carried out on the affected plot, the compensatory planting is normally realised on some other plot (Stejskal 2016, p. 102). If the plot is not directly owned by the party seeking permission to fell trees, § 9 para. 2 ZOPK determines that the owner must give their permission. Each municipality designates plots of land suitable for compensatory plantings.

The nature conservation authorities must provide clear justification for their decision within the framework of their discretionary powers (Stejskal 2016, p. 102). In its judgement of 15 May 2015, the Supreme Administrative Court of the Czech Republic expressed an opinion on the application of the discretionary powers of the administrative organs. In the examined case, 35 pine trees were felled to permit the construction of detached family houses. Regarding suitable compensation, the court determined that the decision on compensatory planting of 145 deciduous trees with a 5-year period of care was within the discretionary powers of the relevant agency (Supreme Administrative Court 2015).

The Ministry of the Environment has instructed the country's Agency for Nature and Landscape Conservation (hereafter termed "Agency") to develop a method to calculate the amount of environmental damage and its compensation (Ministry of the Environment 2014). In the publication "Evaluation of trees outside woodland including the calculation of compensation measures for felled or damaged trees", the Agency has presented a comprehensive method for these calculations (Agency 2013a).

The method consists of the following phases:

1. Gathering of the objective input data: taxon, trunk diameter, tree height, diameter of the crown, physiological vitality of the tree, health, biological value of the taxon, biological value of the biotope.
2. Evaluation of the tree's value—evaluation in a points system determined by the input data, translated into monetary value (the result is an estimation of the value of the environmental loss).
3. Calculation of the compensatory measures—determination of species and number of trees for compensatory planting, points-based evaluation translated into monetary value so that the value of compensation equals the value of the environmental loss.

A software tool is available online for the easy implementation of this method of evaluation and calculation (Agency 2013b).

§ 9 para. 3 of the ZOPK determines that if no compensatory planting is ordered, the party carrying out the tree felling is required to pay financial compensation. This money flows into the municipal budget and is earmarked for the purpose of environmental protection. Concrete preconditions and the amount of financial compensation is supposed to be regulated in a separate law. However, as such a law has not yet been promulgated, this provision cannot be applied (Stejskal 2016, p. 103).

In some cases the application for permission to fell trees is replaced by an obligation of notification. These exceptional cases are specified in § 8 para. 2 ZOPK (e.g. tree felling in a protected zone beside electricity transmission cables or gas pipelines). The nature conservation authority must receive written notification of the felling measure at least 15 days before the scheduled work. It then has the authority to restrict, postpone or forbid the felling insofar as this contradicts the principles of tree conservation. Similarly, prior permission is unnecessary if the condition of trees clearly and directly endangers the life or health of people, or could result in considerable damage (§ 8 para. 4 ZOPK). In such cases documentation of the dangerous condition of the trees must be subsequently submitted to the responsible nature conservation authority within a period of 15 days.

2.2 *Measures to Compensate the Negative Impact of a Plan or Project Within Natura 2000 Sites (§ 45i Paras. 9–11 ZOPK)*

A further obligation to compensate biodiversity losses arises from the Habitats Directive (92/43/EEC) in relation to the Habitats Directive Assessment ("HDA"). Art. 6 (3) and (4) of the Habitats Directive lay down the procedure that has to be followed when any new development is planned which might significantly affect a Natura 2000 site. Compensatory measures shall apply when, due to overriding public interest, a plan or project is to proceed despite having negative impacts. These measures should ensure that the overall coherence of Natura 2000 is protected.

Obligations arising from Art. 6 (3) and (4) of the Habitats Directive are implemented in § 45h and § 45i ZOPK. The procedure follows the steps of the more comprehensive Environmental Impact Assessment (EIA) or Strategic Environmental Assessment (SEA), and uses some of their procedural instruments. A person independent of the nature conservation authority and with special authorization from the Ministry of the Environment carries out the HDA statement.

In general, the provisions regarding the HDA have been carefully transposed into Czech law. However, there is one problematic regulation: *"If the HDA statement contains compensatory measures, a nature conservation authority may only stipulate these compensation measures"* (§ 45i (11) ZOPK). This rule is rather restrictive, preventing the nature conservation authorities from ordering compensatory measures that are discovered after completion of the HDA (Prchalová 2010, p. 151). Therefore, in its guidance document, the Ministry of the Environment advises that compensatory measures should not be determined in the HDA itself, but rather should be left for nature conservation authorities to decide at later stages (Ministry of the Environment 2011, p. 80). The guidance document also introduces a table of all species and habitats, detailing potential compensatory measures and the form they can take (Ministry of the Environment 2011, pp. 82–94).

As specified in Art. 6 (3) and (4) of the Habitats Directive, compensatory measures should be resorted to only when, despite a negative assessment of the impact on a site, there exist compelling reasons of public interest for the implementation of a plan or project. The legal provision of compensatory measures is only rarely invoked in the Czech Republic. Investors frequently withdraw, revise or relocate projects to non-Natura 2000 sites when a negative effect on Natura 2000 has been confirmed, as they view compensatory measures to be too costly and time-consuming (Bejček 2011, p. 13–14). In this respect, the HDA serves to protect Natura 2000 sites in a preventive way.

One case in which compensatory measures have been ordered in the Czech Republic relates to the construction of a bypass in the city of Břeclav in south Moravia. The negative impact of the project was based on the deterioration of habitats in the SCI Soutok–Podluží, in which compensatory measures were therefore ordered. These foresaw the creation of a new habitat of riparian forest and alluvial continental meadow, 2.5 times larger than the damaged area. These measures have already been implemented. The nature conservation authority will now monitor the success of the measures in coming years. The project cannot proceed until there is clear evidence of the success of the compensatory measures.

2.3 Compensatory Measures for Unauthorised Interventions (§ 86 Abs. 2 ZOPK)

The ZOPK also specifies the obligation to provide compensation for unauthorised interventions in nature and the landscape, i.e. those which have not been granted official permission (§ 86 ZOPK). Accordingly, persons who damage, destroy or transform nature and landscapes protected by the ZOPK in some unauthorised way are obliged to restore affected sites to their original condition. If such restoration is impossible or inappropriate, then the party causing the damage is obliged to implement suitable compensatory measures.

These compensatory measures are intended to offset the negative impacts of unauthorised interventions. Compensatory measures are imposed on the party causing the negative impact at the discretion of the nature conservation authorities. Possible measures are, for example, compensatory plantings as well as programmes to increase wildlife or fish stocks (Stejskal 2016, p. 428).

3 Instruments Provided by Other Laws

The obligation to provide compensation is also determined by other laws dealing with the environment apart from the ZOPK.

3.1 Law on the EIA and SEA

Zákon č. 100/2001 Sb., o posuzování vlivů na životní prostředí, "ZOPV"

The obligation to provide compensation is dealt with as part of the Environmental Impact Assessment (EIA) and the Strategic Environmental Assessment (SEA). Thus, Art. 5 para. 3 b) of the EIA Directive (2011/92/EU) determines that the details of a project supplied by the project operator must include a description of measures to avoid, reduce and even to compensate for any major negative impacts. The SEA Directive (2001/42/EG) also regulates for compensatory measures, specifically in Appendix I g): *The environmental report shall encompass measures envisaged to prevent, reduce and offset as fully as possible any significant adverse effects on the environment caused by the plan or programme implementation.*

The European EIA and SEA Directives were incorporated into Czech law through Law No. 100/2001 on Evaluating Environmental Impacts (ZOPV 2001). This law regulates both the EIA (§§ 4–9e ZOPV) as well as the SEA (§§ 10a–10j ZOPV). In both cases mention is made of compensatory measures.

Specifically, operators are obliged to suggest measures to avoid or reduce negative environmental impacts of a project (§ 5 para. 4 ZOPV). If avoidance or at least some significant reduction in negative impacts is impossible, then compensatory measures can also be proposed (Bahýĺová et al. 2015, pp. 39–40). The effectiveness of the proposed measures should be assessed within the framework of the EIA. Furthermore, as part of the SEA, the responsible authority can suggest compensatory measures in their official statement on a plan (§ 10g para. 2 ZOPV).

The binding official statement on a project or a plan represents the end result of the entire evaluation process and is adopted as a strict basis for decisions in subsequent processes (e.g. the adopting of binding land use plans or the issuing of building permits). However, the regulation on compensatory measures is in fact rarely applied in practice.

3.2 *Law on Environmental Damage*

Zákon č. 167/2008 Sb., o předcházení ekologické újmě a o její nápravě, "ZOPEÚ"

A further form of compensation is determined by the Law on Environmental Damage, which is also rooted in European law. The Law No. 167/2008 on the Avoidance and Offsetting of Environmental Damage (ZOPEÚ 2008) transposes the European Directive on Environmental Liability (2004/35/EG) into Czech law. This regulation deals with the cause of an existing case of environmental damage or some impending environmental damage due to an occupational activity that is subject to a strict liability. The governing principles of this regulation are the *polluter pays principle* and the *precautionary principle*. Environmental damage is defined as damage to species and habitats, damage to water and damage to soil.

The party responsible for existing or impending environmental damage has numerous obligations, including the duty of remediation. The remediation of damage to protected species and habitats is regulated in § 10 ZOPEÚ and the accompanying Appendix No. 4. The duty of remediation consists of so-called primary remediation (restoration of the damaged natural resources and/or deterioration in natural functions completely or approximately to their original state), supplementary remediation (additional measures in those cases where primary remediation does not result in complete restoration) and compensatory remediation (compensation of some intermediate loss of natural resources and/or functions before the primary remediation is achieved).

3.3 *Law on Land Use Planning and Building Regulations*

Zákon č. 183/2006 Sb., o územním plánování a stavebním řádu—stavební zákon, "SZ"

Finally, obligations to provide compensation for interventions in nature and landscapes are also anchored in the Law on Land Use Planning and Building Regulations (SZ 2006). Thereby, one of the tasks of land use planning is, in the case of negative impacts of plans and projects, to propose and to implement compensatory measures (cf. § 19 para. 1 m SZ). The compensatory measures are derived from other laws, in particular the Law on the Assessment of Environmental Impacts as well as the Law on Nature and Landscape Conservation (see above).

4 Conclusions

In summary, we can say that while Czech environmental law specifies diverse obligations regarding compensation for interventions in nature and landscape, a general nationwide instrument to offset negative impacts on biodiversity is lacking. A comprehensive method to assess losses in biodiversity only exists within the framework of conservation of trees and bushes outside woodland. Here the form and extent of compensatory measures (generally replacement plantings) are at the discretion of the responsible administrative authority.

The obligation to restore sites to their former condition or to compensate for unjustified interventions (these are called *corrective measures* in Czech law) strictly issues from tort liability. In other words, such duties arise as secondary obligations following the violation of some relevant environmental protection law. The general regulation of tort liabilities for interventions in nature and landscape is rooted in the Law on Nature and Landscape Conservation. The Law on the Avoidance of Environmental Damage contains a special provision on environmental damage caused by occupational activities.

Obligations to provide compensation originating in European law (particularly within the frame of EIA/SEA and the Habitats Directive Assessment) are rarely applied in the Czech Republic. In the case of a negative result of the Habitats Directive Assessment, investors tend to alter the technical characteristics or the location of projects rather than implementing expensive compensatory measures in Natura 2000 sites. This should be welcomed as encouraging fewer interventions in such protected areas and hence protecting the Natura 2000 network. However, interventions will then be steered towards other areas in which, apart from the protection of trees and bushes, no other duties of compensation exist.

References

Agency (2013a) Kolařík J et al 2013 Oceňování dřevin rostoucích mimo les včetně výpočtu kompenzačních opatření za kácené nebo poškozené dřeviny. Metodika AOPK ČR, Praha (Evaluation of trees outside woodland including the calculation of compensation measures for felled or damaged trees)

Agency (2013b) Safe trees 2013 software tool for evaluation and calculation. http://www.ocenovanidrevin.nature.cz/

Bahýlová L, Kocourek T, Vomáčka V (2015) Zákon o posuzování vlivů na životní prostředí. Komentář. 1. vyd. Praha: C. H. Beck, s.r.o., 2015 (Commentary on the Law on EIA and SEA)

Bejček P (2011) Kompenzační opatření podle článku 6.4 směrnice o stanovištích. Ochrana přírody 3/2011 (Compensatory Measures under Art. 6 para. 4 of the Habitats Directive)

Ministry of the Environment (2011) Příručka k hodnocení významnosti vlivů na předměty ochrany lokalit soustavy Natura 2000 (Guidelines on Habitats Directive Assessment)

Ministry of the Environment (2014) Metodický pokyn odboru obecné ochrany přírody a krajiny Ministerstva životního prostředí ze dne 1. 8. 2014 k aplikaci § 8 a § 9 zákona č. 114/1992 Sb., o ochraně přírody a krajiny, ve znění pozdějších předpisů upravujících povolení ke kácení dřevin rostoucích mimo les a náhradní výsadbu a odvody, ZP29/2014 (Methodological guidelines on the application of § 8 and § 9 of Law No. 114/1992 on Nature and Landscape Conservation, which regulates the authorisation of tree and bushes felling outside woodland as well as replacement plantings)

Prchalová J (2010) Zákon o ochraně přírody a krajiny a Natura 2000: úplné znění zákona s komentářem, judikaturou a prováděcími předpisy. 2. aktualizované a rozšířené vyd., Praha: Linde (Law on Nature and Landscape Conservation and Natura 2000 – Commentary, Case Law and Provisions for Implementation)

Stejskal V (2016) Zákon o ochraně přírody a krajiny. Komentář. 1. vyd. Praha: Wolters Kluwer ČR, a.s., 2016 (Commentary on the Law on Nature and Landscape Conservation)

Supreme Administrative Court (2015) Judgement of the Supreme Administrative Court of the Czech Republic from 15.05.2015 in the Case 7 As 280/2014–72

SZ (2006) Zákon č. 183/2006 Sb., o územním plánování a stavebním řádu – stavební zákon (Law on Land Use Planning and Building Regulations)

ZOPEÚ (2008) Zákon č. 167/2008 Sb., o předcházení ekologické újmě a o její nápravě *(Law on the Avoidance and Compensation of Environmental Damage)*

ZOPK (1992) Zákon č. 114/1992 Sb., o ochraně přírody a krajiny (Law on Nature and Landscape Conservation)

ZOPV (2001) Zákon č. 100/2001 Sb., o posuzování vlivů na životní prostředí (Law on the Assessment of Environmental Impacts)

France

Anne-Charlotte Vaissière, Fabien Quétier, Adeline Bas, Coralie Calvet, Stéphanie Gaucherand, Julien Hay, Céline Jacob, Charlène Kermagoret, Harold Levrel, Aurore Malapert, Sylvain Pioch, and Pierre Scemama

1 Theory in France

1.1 Legal Background in France

1.1.1 Existing Laws and Rules

The country's first nature protection law to introduce environmental impact assessments and establish the mitigation hierarchy was passed in 1976. The mitigation hierarchy, however, was not properly implemented until a number of recent reforms were introduced (to transpose EU Directives as well as to reflect society's increasing concern for

A.-C. Vaissière (✉)
Ecologie Systématique Evolution, CNRS, AgroParis Tech, Université Paris-Sud, Université Paris-Saclay, Orsay, France
e-mail: anne-charlotte.vaissiere@u-psud.fr

F. Quétier
BIOTOPE, Mèze, France
e-mail: fquetier@biotope.fr

A. Bas · J. Hay · C. Jacob · P. Scemama
AMURE, Université de Brest, IFREMER, CNRS, IUEM / AMURE Unité d'Economie Maritime, IFREMER, Université de Brest, CNRS, IUEM, Plouzané, France
e-mail: adeline.bas@univ-brest.fr; julien.hay@univ-brest.fr; celine.jacob@ifremer.fr; pierre.scemama@ifremer.fr

C. Calvet
CEE-M, CNRS, Université de Montpellier, INRA, SupAgro, Université Paul Valéry Montpellier, Montpellier, France
e-mail: coralie.calvet1@gmail.com

S. Gaucherand
UR EMGR, Université Grenoble Alpes, IRSTEA, St-Martin-d'Hères, France
e-mail: stephanie.gaucherand@irstea.fr

W. Wende et al. (eds.), *Biodiversity Offsets*,
https://doi.org/10.1007/978-3-319-72581-9_7

biodiversity), with offsets being increasingly required in recent years. The issue was further highlighted by conflicts around several large projects which required the establishment of offsets to deal with impacts on wetlands and endangered species. The applicable definition of offsets in France is found in article R.122-14-III of the French Environmental Code. These are measures that '*aim to compensate the significant negative effects, direct or indirect, of the project that could not otherwise be avoided or sufficiently reduced. They should be implemented on the damaged site or in proximity to it so as to ensure its functionality through time. They must ensure that the environmental quality of habitats is globally maintained or, if possible, enhanced*' [our translation]. While there is no distinction in French between "compensation" and "offsets", the reference to maintaining habitat quality is equivalent to the no-net-loss goal.

Offsets are required for the approval of various types of projects, plans and programs through different procedures. These include Environmental Impact Assessments (EIA), also for high risk projects (as per SEVESO Directives), and Strategic Environmental Assessments (SEA), but biodiversity offsets are most frequently demanded in the context of permit applications for projects in or near Natura 2000 sites (Appropriate Assessments under the Habitats Directive) and projects that could impact protected species and aquatic habitats (Quétier et al. 2014). Recent changes to the relevant regulations[1] opens up the possibility for a developer to submit a single environmental permit application under several regulatory procedures. The requirement for biodiversity offsets currently applies to:

- Aquatic habitats and wetlands, for which watershed management plans (*Schémas Directeur d'Aménagement et de Gestion des Eaux—SDAGE)* require offsets for wetland degradation and destruction as well as any other impact which would

[1] https://www.legifrance.gouv.fr/eli/decret/2017/1/26/DEVP1621458D/jo/texte and https://www.legifrance.gouv.fr/eli/ordonnance/2017/1/26/DEVP1621456R/jo/texte

C. Kermagoret
Département des sciences naturelles, Institut des sciences de la forêt tempérée, Université du Québec en Outaouais, Gatineau, Canada
e-mail: charlene.kermagoret@gmx.fr

H. Levrel
CIRED, AgroParisTech, Nogent-sur-Marne, France
e-mail: harold.levrel@agroparistech.fr

A. Malapert
BIOTOPE Chine – Feilan, Beijing, China
e-mail: amalapert@biotope.fr

S. Pioch
CEFE, Université Paul-Valéry Montpellier 3, CNRS, Université de Montpellier, EPHE, Montpellier, France
e-mail: sylvain.pioch@univ-montp3.fr

undermine the Good Ecological Status objectives of the Water Framework Directive (WFD). This is usually done through an area-based ratio of 1.5 or 2 ha restored area for 1 ha destroyed area, and/or through restoration that ensures no net loss of wetland functions (see Sect. 1.2.2 on metrics).

- Similarly, impacts on marine biodiversity can give rise to offsets in the context of Marine management plans (*Plans d'Action pour le Milieu Marin—PAMM*) as per the Marine Strategy Framework Directive.
- Habitats protected under the Habitats Directive, including those that justified the designation of Natura 2000 sites, for which specific permits must be sought as per Article 6 of the Habitats Directive.
- Protected species, and their habitats (including but not limited to those that justify the designation of Natura 2000 sites), for which derogations must be sought for projects that could affect their conservation status as per Articles 14–16 of the Habitats Directive, but with a broader list of target species. This is currently the main driver of offsets in France.
- Forests, for which clearing permits are required under the Forest Code. Here offsets in the form of afforestation are often sought by the French Forestry Agency (*Office National des Forêts*). A 2014 reform of the Forest Code now allows financial compensation through payments to a trust fund (*Fonds stratégique de la forêt et du bois—FSFB)*. It should be noted that impacts on forests can generate mitigation measures and offsets due to their value as habitats for protected species.
- Wildlife corridors identified in regional level plans (*Schéma Régional de Cohérence Ecologique—SRCE*) as "*green and blue veins*" could generate offsets if they are impacted. However, the SRCE has only recently been established, and no such cases have yet been documented. However, corridors relevant to protected species are often considered in the corresponding derogation applications.
- Ecosystem services are not clearly referred to in environmental regulations but are sometimes investigated in the context of EIA. They do not give rise to offsets but are instead considered under public information and participation procedures where the concerns of stakeholders and sectoral interests can be heard (see Jacob et al. 2016b for a discussion).

A major impetus for biodiversity offsets was the introduction in 2007 of derogations for impacts on protected species and their habitats (as per Articles 14–16 of the Habitats Directive), with a national consultative body (*Conseil National de Protection de la Nature*) providing an informed opinion on the derogation. In 2012 approval procedures were modified to require that mitigation and offset measures be explicitly included in the permits themselves and not just in the EIA and SEA documents. This makes them legally binding (Quétier et al. 2014). Furthermore, the scope of EIA and SEA was broadened and enforcement restructured, supposedly establishing clearer sanctions if measures were not put in place by permit holders.

Given these changes, the French Ministry for the Environment provided guidance on applying the mitigation hierarchy and offsets in 2012 (MEDTL 2012) and 2013 (MEDDE 2013). Although not actual legislation, these guidance documents are the main references available for developers and regulatory authorities, EIA practitioners, and courts of law. Section 1.2.1 explains in greater detail how the mitigation hierarchy is implemented in France and which limits it faces.

1.1.2 Evolution of the French Legal Context

In recent years France has introduced a number of changes to its regulations covering impacts on biodiversity. In particular, a national consultative process on the environment initiated in 2007, the "*Grenelle de l'Environnement*", has ensured political traction and stakeholder involvement for reforms that are still being implemented today, including the EIA and SEA reforms of 2012 and 2013. The French legal context was transformed by the new biodiversity law[2] adopted in 2016 after having been reviewed and amended in parliament for more than 2 years. One of the objectives of this new law was to improve the application of the mitigation hierarchy. In line with the official guidance of 2012 and 2013, the law laid down the objectives of "no net loss" and "net gain". The law also established the option for developers to select from three mechanisms to implement the biodiversity offsetting requirements: the permittee responsible system and two third-party systems, namely the mitigation bank system (called *site naturel de compensation*) and the "offset operator".

Although developers remain liable for the success of their mitigation measures, the law seeks to clarify the rights and duties of third-parties contracted to implement a permittee's offset, especially when such an "offset operator" is not the owner of the land supporting the offset.

The law also introduces a type of conservation easement (called *obligation réelle environnementale)* that can be applied to guarantee that the land will be used for conservation. However, there is no requirement for the conservation easement to last in perpetuity.

To track offsets and facilitate monitoring and enforcement, several local regulatory authorities have created databases and GIS tools. Some of these databases are publicly accessible. However, many authorities have faced difficulties in gathering the necessary information from developers for past permits, and databases have not been harmonized (Bull et al. 2018). In fact, the law requires the creation of a national, publicly-available and georeferenced database on offsets.

The opening articles of the law mention the concept of ecosystem services within the definition of the mitigation hierarchy. Under the current wording, it seems that ecosystem services only have to be considered during the avoidance and reduction steps of the mitigation hierarchy.

---

[2]https://www.legifrance.gouv.fr/eli/loi/2016/8/8/DEVL1400720L/jo/texte

1.1.3 Offsetting Accidental Impacts

The remainder of this chapter will focus on biodiversity offsets resulting from *ex ante* assessments in the context of permitted impacts on biodiversity. In this section, we provide essential information and important progress that has been made in offsetting accidental impacts.

In 2008 France transposed the EU's Environmental Liability Directive (ELD) into national law. The aim of the Directive is to prevent and remedy environmental damage by requiring primary, compensatory and complementary measures to be taken to offset damage to protected natural habitats and species. It also suggests specific equivalency methods to be favoured when designing and sizing compensatory restoration (DCGSD 2011), specifically dedicated software such as Visual HEA 2.6, which was developed to support implementation. To date, however, the ELD has yet to be applied in France (Mudgal et al. 2013; Martin 2014).

The capacity to enforce offsets is also enhanced by another article of the 2016 law which enables greater scrutiny of developers. The Erika oil spill of 1999 led the highest civil court to determine that "ecological damage" (understood as damage to the environment) may be compensated. Following this landmark ruling, several publications by dedicated academic and institutional working groups suggested various amendments to the French Civil Code in order to improve the remediation of damage to natural resources (Neyret and Martin 2012; Le Club des Juristes 2012; Groupe de travail sur le préjudice écologique 2013). These amendments have been included in the recently passed law on biodiversity, which also indicates that compensation of damages should be done in kind through conservation and restoration activities. It is expected that this introduction of ecological damage into the Civil Code will address some of the obstacles to ELD implementation in France. However, it is still not clear if and how ecological damage in relation to the faulty implementation of biodiversity offsets can be legally pursued.

1.2 Methodological Background in France

1.2.1 Mitigation Hierarchy in France

In spite of recent efforts to establish the mitigation hierarchy more soundly in France, its current application remains highly heterogeneous, depending on where projects are located (regional approval authorities vary in their approach) and the sectors concerned, e.g. large linear infrastructure vs. housing or wind energy projects (developers, their service providers and permitting authorities follow sector-specific practices). Until recommendations from national-level guidance are followed more widely, the burden of designing and implementing adequate offsets lies with local and regional approval authorities, who are under pressure from local elected officials keen to promote development as well as developers themselves. While the lack of an independent environmental authority has recently led to the creation of regional environmental authorities, these appear ineffectively organized and funded. Consequently, and in spite of the increasing demand for offsets, we find a highly variable and often ineffective project-by-project approach to biodiversity offsets, typically with minimal requirements. To improve matters, a number of regional and national initiatives are striving to provide a clearer and more practical framework for the harmonization of mitigation practices. In the following we examine some of the main stumbling blocks towards an improved and harmonized application of the mitigation hierarchy in France.

Complex and Unclear Procedures
The impression amongst developers is generally of long, complex, heterogeneous and unclear approval procedures, in particular when their projects are subject to different procedures for which administrative and technical overlaps are not well defined, e.g. projects that impact wetlands harboring protected species are subject to two separate approval procedures in addition to EIA. Approval authorities are critical of the low quality of permit applications, which, however, are large in number. This slows the review process while the growing scope of permits and the legal risk they carry drive developers to submit ever more applications. In particular, the distinction between avoidance, reduction (minimization) and compensation/offsets is often misunderstood. Approval authorities have limited manpower to review applications or subsequently monitor and enforce the implementation of mitigation measures. A key concern is that methods used to design and determine the extent of offsets are often unexplained (Bigard et al. 2017; and see Sect. 1.2.2), a situation that often serves to delay reviews (Jacob and Pioch 2014).

A Lack of Forethought in Land-Use Planning
Another obstacle to the adequate review of permit applications is the lack (at least in part) of long-term visions or objectives for biodiversity at the local or regional scale beyond the assessment of individual projects. This makes cumulative impact assessment difficult, and prevents the inclusion of offsets in current/future conservation or restoration plans and policies. If greater consideration were given to offsets over the longer term, this would facilitate their aggregation into the best locations for long-term effectiveness through integrated spatial planning, as well as allow better forecasting of the need for land or capacity for implementing offsets (Vaissière et al. 2016). Moreover, objectives set at the landscape level would certainly enable more consistent assessments of the significance of impacts (Briggs and Hudson 2013). There is also considerable scope for improving SEA practice, as biodiversity is often a low priority issue when developing local land-use plans, and is rarely considered in SEA with the same level of attention as in EIA. Also, some plans are now quite old and in need of revision; they were designed at a time when the level of concern for biodiversity was much lower.

Although recent changes to the French EIA procedures ensure that other plans or projects must now be carefully considered for a project's approval, the issue remains that cumulative impacts of several projects at a given spatial scale are largely addressed through a "first come, first served" approval process. This sets the stage for the progressive lowering of thresholds beyond which additional impacts can no longer be considered negligible (i.e. "death by a thousand cuts"). Furthermore, there is no recommendation to ensure compatibility between the data or methods used for assessing impacts across projects. It can also prove difficult to access data from third party projects. Unfortunately, developers are given no assistance in this matter.

Clearly, improved coordination is required to overcome the project-by-project approach to mitigating and offsetting impacts on biodiversity.

Little Anticipation of the Cost of Mitigation Measures
Due to scanty forethought in land-use planning, the cost of mitigation and offsetting is often poorly budgeted by project proponents, who "discover" biodiversity issues long after they have secured funding. This can be explained by the fact that the specific approval procedures required for protected species—which, as explained above, trigger most offsets—are initiated *after* the EIA process and the associated public hearings. Developers need to be made aware of biodiversity-related risks earlier in their investment decisions. This would require more precise mapping of risks, e.g. data on the presence and numbers of protected species as well as prior analysis of key habitats in and outside protected areas. France has a system for this: the ZNIEFF network (*Zones Naturelles d'Intérêt Ecologique, Faunistique et Floristique*), which specifies sites of special interest. However, this appears unsuited to dealing with project-level risks outside the actual network. While a previous national initiative to develop communal biodiversity atlases proved unsuccessful, it was relaunched as part of an updated national biodiversity strategy. Because mitigation and offsets are insufficiently budgeted, they are not incorporated into cost-benefit analyses of projects, even when such analyses are required (e.g. for large public infrastructure) or when the social and economic justification of the project is called into question. In fact, an important concern in the application of the mitigation hierarchy in France is the weakness of the justifications of public interest. Approval authorities which consider permit applications concerning biodiversity often do not review the justifications put forward by applicants, which are often simple and unsubstantiated statements in terms of direct and indirect job creation.

Lack of a Shared Methodological Framework
In France there are no standardized approaches for assessing equivalence and comparing biodiversity losses and gains, which are necessary to demonstrate that no net loss can be achieved or to adequately weigh the risks of failing to achieve the no net loss objective. The only exception is for impacts on the Common Hamster (*Cricetus cricetus*), for which a standard assessment method is accepted under Ministerial Order DEVL1231144A of August 2012. The establishment of rigorous approaches to defining loss-gain metrics would help to harmonize mitigation practice in France, especially as offsets are required for recurring impacts on widespread habitats and protected species. For these, more standardized metrics at landscape scale would be useful. To date, only one exemplary method has been developed for wetland functions (Gayet et al. 2016). Section 1.2.2 goes into more details on metrics in France.

Unsatisfying Management of Spatio-Temporal Aspects of Offsets
In terms of actual implementation, the duration of offsets, in particular, has been found to vary greatly, as do the legal and financial guarantees that developers have to

provide for that duration. The new law on biodiversity requires that offsets be applied for the duration of the impacts. Furthermore, although guidance indicates that offsets should be in place before impacts occur, this is still rarely the case, underlining the importance of drawing up viable plans for offset implementation. This usually means identifying partners, and multiple locations or options to achieve the required ecological outcomes.

Access to land has to be anticipated, either for purchase or for contracting with landowners and land-users (e.g. in an agricultural context—see Calvet et al. 2017). Appropriate guidance on the location of offsets is essential to ensure that developers do not systematically reduce the cost of offsetting by targeting cheaper land, often far from areas under development pressure, and with no consideration of the consequences in terms of no net loss goal or ecological feasibility. Under the 2016 biodiversity law, the French agency for biodiversity is supposed to identify the public and 'abandoned' lands that would be suitable for offset implementation. Requiring that offsets be located close to impacts does not, however, automatically improve ecological outcomes. While the law on biodiversity usefully recommends "functional proximity" between impacts and offsets, this should probably be better defined, and is in contradiction with the expressed preference for on-site offsetting.

Weak Monitoring and Control

Although mandatory after the EIA reform of 2012, monitoring and reporting of offset implementation are often weakly realized. Indeed, little guidance is available on the design of monitoring schemes (e.g. indicators, protocols, frequency and duration) or on procedures for adapting offsets to monitoring results. While monitoring and reporting is the responsibility of developers, the resulting procedures do not always target the biodiversity features which generated the offset. Furthermore, these tasks tend to be subcontracted to third parties, often those actually implementing the offset. To date, no mechanisms have been set up in France to ensure offset performance is verified by competent third parties, although regulatory bodies can require this on a case-by-case basis. Enforcement of offset implementation (if not performance) is, however, on the increase, especially in relation to waterways and wetlands. Developers increasingly see offset implementation as an important risk factor in their operations.

Frequent Confusion Between Biodiversity Offsets and Other Accompanying Measures

In practice, biodiversity offsetting is often considered within a broader set of measures aiming to improve the likelihood that the goal of ecological no-net-loss is met but also to maintain principles of social justice for the impacted communities and, above all, to increase the social acceptability of a project (Kermagoret et al. 2014; Gobert 2010). In France measures not considered as forms of avoidance, minimization or offsets are generally called "accompanying measures". Their UK equivalent includes "community benefits". They are voluntarily implemented by developers to support many development projects by funding or investing in actions for local communities (Bristow et al. 2012; Walker et al. 2014). In this context, ecological restoration actions may be designed to address the social issues of ecological impacts (Gobert 2010) and allow developers to circumvent the methodological difficulties related to the no-net-loss goal for impacted biodiversity features. This generates "socio-environmental" compensation, sometimes by redefining the no-net-loss on a more anthropocentric basis. To this end ecosystem service approaches are sometimes adopted. As well as creating confusion in the goals and definitions of the mitigation hierarchy, this raises a technical challenge of ensuring that no-net-loss goals can still be met when including these social considerations (Jacob et al. 2016b).

1.2.2 Metrics in France

A Focus on Area

Since the first mention of the mitigation hierarchy in French law in 1976, the metrics used to demonstrate equivalence have basically remained the same: developers, approval authorities and stakeholders have focused on area-based ratios or multipliers (UICN France 2011, Quétier and Lavorel 2011, Regnery et al. 2013). Divergent approaches tailored to the specific regulations that trigger biodiversity offsets (i.e. for forests, wetlands, Natura 2000 habitats, protected species or corridors) are only now starting to emerge.

For forested lands the ratio is usually 1:1, but can reach 5:1 depending on the ecological and social importance of the forest as agreed by the developer and its stakeholders (notably the National Forest Office which manages France's public forests). For wetlands, regulations (SDAGE) can require the restoration of twice (or 1.5 times) the surface of wetlands impacted by a proposed development. In some watersheds a 1:1 ratio is specified if the developer can demonstrate that this is sufficient to maintain all wetland functions of the impacted watershed (De Billy et al. 2015). The National Office for Water and Aquatic Habitats (ONEMA) recently published a methodology to assess such functions (Gayet et al. 2016). This may shift offsetting for wetland impacts from the currently focus on area (regardless of quality) to a more quantitative analysis of losses and gains of wetland functions. A preliminary analysis by Vaissière et al. (2016) has shown that this can enable more effective targeting of restoration efforts at the landscape scale. For protected species, area ratios can reach 10:1 (Regnery et al. 2013), for example in the case the habitat of the endangered Hermann's tortoise (*Testudo hermanni*) in Provence (DREAL PACA 2009). The specific ratio is defined on a case-by-case basis, mainly on the basis of the conservation status of the impacted species. Progressively, this is leading to species-specific ratios being set as derogations, which are regularly granted for impacts on individual species. Unfortunately, this ignores the actual level of impact suffered by the species at a given location, e.g. on the basis of habitat quality. For plant species, the number of individuals directly destroyed by a development is also considered (Gaucherand et al. 2013).

In principle, whichever procedure is concerned, the 2012 French guidance aims for strong levels of sustainability through highly targeted like-for-like equivalence (e.g. species for species at a local level under the derogation procedures) (Quétier and Lavorel 2011).

The Limits of Area-Based Metrics
Using area to assess impacts and size compensation makes sense. Indeed, species diversity and viability are related to the available area of a habitat (Enviroscop et al. 2010 pour le Ministère de l'Ecologie). However, surface area is not a sufficient indicator of impact. Similar habitats can vary in their ability to perform different functions depending on their characteristics such as age, species composition, soil or topography as well as the position in the landscape (Quétier and Lavorel 2011). Also, restored or artificially created habitats tend to be less functional than pristine habitats (Aulert et al. 2009; Moreno-Mateos et al. 2012). Finally, the ratios used to determine the extent of compensation include a risk factor to account for the likelihood of failure during the restoration process. This risk factor is rarely based on scientific evidence. Area ratios are generally determined through negotiation between developers and approval authorities against a background of stakeholder pressure, particularly from other land-users such as farmers and hunters. Developers are often keen to favour land purchase opportunities which present themselves even if these are not fully suitable for the intended offsetcompensation ratios, there is no guarantee of achieving a no net loss of biodiversity if the surface area of a given habitat remains the only metric used to determine equivalence as well as the design and extent of offsets (Quétier and Lavorel 2011). This approach is clearly inefficient both economically and ecologically.

Alternatives to Area-Based Metrics
In their reviews, French researchers have already suggested various ways to improve the metrics used to assess equivalence (Barnaud and Coïc 2010; Quétier and Lavorel 2011; De Marsily et al. 2013; Regnery et al. 2013; Quétier et al. 2014; de Billy et al. 2015; Levrel et al. 2015; Vaissière et al. 2016, among others). The two main approaches to date are: (a) multi-criteria approaches to calculating offset metrics, and (b) loss-gain approaches. The latter are based on Equivalency Analysis, which aims for a quantitative analysis of no net loss, and include the definition of biodiversity metrics and exchange rules. Typically, metrics are related to habitat quality (e.g. for a species or Natura 2000 habitat) or ecosystem functions (e.g. for wetlands), and can take into account ecosystem dynamics (e.g. ecological trajectories of the impacted and the compensatory sites) as well as spatial issues such as connectivity, location of the impacted and offset site in relation to one another. Exchange rules consider timing, risk, etc. to determine whether an offset is acceptable. Such methods have been used in large infrastructure projects, where developers have been confronted with complex sets of biodiversity features that require avoidance, mitigation and offsetting. Specialized consulting firms and public research institutes are often involved in developing tailored Equivalency Analysis frameworks. This was the adopted method for developing a standardized rapid assessment method for wetland functions (Gayet et al. 2016). However, it appears that for simpler, routine, applications of the mitigation hierarchy, area-based ratios remain the preferred approach. Multi-criteria methods have been put together to rationalize the use of area-based ratios when protected species are impacted (criteria include the local conservation status of the species, the ability of the species to colonize, the area of the impact and the number of individuals impacted, the nature of the impact, the

efficiency of the restoration process, the distance between impacted and compensatory site, etc.). However, these methods do not consider criteria related to ecosystem functions, and do not offer a transparent loss-gain framework as recommended by international good practice (Quétier and Lavorel 2011).

Currently, a large number of methods and approaches are being investigated to develop offset metrics that go beyond surface area and arbitrary ratios while still observing French regulations. The risk is that we end up with a proliferation of methods with little feedback and no consensus among developers, consulting companies and the environmental authority. Furthermore, the report by UICN-France (2011) suggests that offset metrics should find a balance between limiting substitution and establishing a currency that is sufficiently fungible to facilitate exchange. This implies the use of like-for-unlike equivalence. In Quétier et al. (2014), the authors suggest that when strict equivalence (like-for-like) between losses and gains cannot be implemented and yet a project will still go ahead for overriding reasons of public interest, impacts on a less valuable type of biodiversity could be offset through actions in favour of a more valuable type of biodiversity. However, specific metrics to facilitate such exchange will only be developed if the French regulations allow for this option, which is currently not the case.

1.2.3 New Habitat Banking in France

While French legislation recently provided a framework for implementing biodiversity offsets through habitat banking, this remains at an experimental (pilot) stage. The first example, initiated in 2008, is presented in Sect. 2.2. Three years later, in 2011, the French government launched a call for proposals for other experimental habitat banks. Those that were approved show great diversity in terms of targeted ecosystems and institutional arrangements. Information regarding these experiments is publically available.[3] While the government wishes to encourage a bottom-up approach in order to test different institutional arrangements for habitat banking, at the same time it gave proponents a roadmap to assure the coherence of the proposed projects with biodiversity offset requirements (e.g. ecological equivalence, anticipation, additionality, transparency of the information regarding the trading of compensatory units, long-term commitments to biodiversity outcomes). The experimental projects were supposed to be compatible with applicable legislation, which did not mention habitat banking until the passing of the 2016 law. Decrees as well as an order on implementing the 2016 law provide clearer if still incomplete rules on the establishment of habitat banks in France (Vaissière et al. 2017).

[3]http://www.developpement-durable.gouv.fr/eviter-reduire-et-compenser-impacts-sur-lenvironnement#e4

2 The Practice of Biodiversity Offsetting in France

We have selected three illustrations of biodiversity offsetting practice in France: two are examples of case-by-case compensation (an older and a more recent case) while the third is the country's only fully established habitat bank. The final section details the case of marine biodiversity offsetting, which is particularly weakly developed.

2.1 Case Study in France: Case-by-Case Compensation

2.1.1 Case-by-Case Compensation for a Port Expansion Project

In 1995 the Port of Le Havre, located in the lower Seine estuary (English Channel), began discussing the construction of "Port 2000", a new facility dedicated to container ships with around 4.5 km of new docks and 5 km of dykes. It was decided to balance this downstream extension by a series of environmental measures that encompassed both regulatory and voluntary compensatory measures, including an expansion of a protected area and the creation of new habitat for shorebirds. A total budget of 46 million euros was set aside to realize these environmental measures, representing just over 4% of the total cost of the port construction project, estimated at 1088 million euros.

In this section we focus on one important compensatory measure, i.e. the creation of roosting habitat on a dune as well as two accompanying measures, i.e. the project to create a meander in the estuary as a way to rehabilitate the mudflat habitat and the construction of an islet resting place for birds at the outlet of the estuary (Fig. 1). Compared to compensatory or offset measures, "accompanying measures can (...) be defined to improve the efficacy of or give additional safeguards to the environmental success of offset measures. They can also target socioeconomic activities" (Jacob et al. 2016b, p. 96).

The Seine estuary is an important site for shorebirds due to its location on the Western European migration route and given the richness and diversity of its natural or semi-natural habitats, notably fresh and saltwater marshes and tidal habitats. As the Port 2000 extension would destroy an area used by shorebirds as a resting place at high tide, authorities in charge of project approval demanded that compensation be provided. The developer suggested restoring an equivalent resting area on a nearby dune habitat. This measure was engineered to maximize the "production" of biodiversity by attempting to take account of the largest possible number of environmental variables. This approach engendered a morass of technological problems and eventually failed. Instead, the management of the area took a more comprehensive direction based on a better understanding of local functions. The new aim was to implement more suitable measures such as extensive grazing or manual management of water levels as well as the integration of other measures in the surrounding area.

Fig. 1 Aerial view of Port 2000 and the environmental measures accompanying the project (Image: Google Earth—Cnes/Spot Image 2013)

For example, it was decided to forbid bird hunting near the area in order to limit disturbances to resting shorebirds.

The creation of an islet in the estuary was intended to provide new resting areas for shorebirds. Although many different stakeholders (with diverse motives) were involved in its design, this plan faced a barrage of criticism due to the high cost of 8 million euros, the fact that an "artificial" habitat was used to compensate the loss of a "natural" habitat, and the fear that the islet could affect the hydro-sedimentary dynamics of the estuary and consequently damage popular recreational beaches downstream from the estuary (in particular Deauville) over the long term. In order to satisfy all the various demands, the objectives of the islet project became more numerous and complex. This has resulted in a sub-optimal result: while the expected species are now observed on the islet, the numbers are below expectations. Hence, the islet fulfills its function as a resting area with limited ecological efficacy (Aulert et al. 2009). The islet is increasingly presented as a compensatory measure to overcome the failings of the dune restoration and conservation measures, in contradiction to initial commitments that this would be an additional compensatory measure.

Mudflats play a very important role in the functioning of the estuary as they constitute an important benthic zone supporting a trophic chain through which both fish (at high tide) and shorebirds (at low tide) are fed. The various forces of urbanization, the construction of embankments and the growth in coastal vegetation have gradually eaten away at the mudflats, reducing their extent by 20 ha annually between 1980 and 2000. An ambitious rehabilitation project involving the creation of a new meander under the Normandy Bridge was proposed to slow this shrinkage. The preliminary phase was to establish a model of the potential hydrodynamic effects of different scenarios with a view to forming 100 ha of functional mudflats. With a budget of 23 million euros, the implemented scenario took 12 months of construction work and involved the dredging of 1,800,000 m^3 of silt and sand to create a meander of around 2800 m in length and 100 m in width. To date, the ecological objective has not been reached: Only 60 ha of functional mudflats have appeared and not in the intended area. While it was decided to make some minor adjustments, the main reaction has been to wait and see how the situation will evolve. There is a considerable risk that the cost of any future upgrading or management measures will exceed available budgets as these are exhausted to fund monitoring.

Overall, none of the three measures achieved their intended goals, leading to a net loss of some biodiversity features as well as a deficit in terms of ecosystem services (Scemama and Levrel 2016).

This outcome can be best understood if we carefully consider the complex situation of the Seine estuary. Here the Port of Le Havre plays a central economic role. In 2010 no fewer than 22.5% of jobs in Le Havre were linked to the industrial/port complex (Camesella et al. 2013). There is strong competition with ports in Belgium and the Netherlands to attract traffic. As a result, the Port 2000 extension enjoyed strong political support. The numerous Natura 2000 sites in the estuary were generally perceived as obstacles to this ambition, imposed by external actors. The story of the creation of the National Nature Reserve clearly illustrates the tensions between development and conservation in the estuary. An initial reserve of 1300 ha was created in 1988 as a compensatory measure for the construction of the Normandy Bridge. In 1990, following legal action by conservation NGOs, the reserve was upgraded to the status of a nature reserve, and extended to 3700 ha. Finally, the size of the reserve was increased to 8500 ha as part of the compensation package for the Port 2000 extension. The reserve is the focus of a number of conflicts. One prominent group in such disputes is the 1800–2000 hunters active in the estuary, who exert a considerable impact on shorebird populations and are vocal in expressing their rights to locally elected officials. Farmers and fishermen often align themselves with hunters in opposing environmental organizations and goals.

In order to increase public acceptance of Port 2000, an open debate consisting of 42 meetings was organized between late 1997 and early 1998. These discussions shaped the design of environmental measures, with many stakeholders expressing dissatisfaction at the succession of development projects and the lack of consideration given to their impacts on the estuary. Many perceived the port's compensation measures as attempts to rectify the failure to sufficiently compensate for past impacts, not just those of the Port 2000 extension. Over time, these measures share a common fate: They are attacked by environmental parties both for the poor results and for the form of implementation. The measures are seen to do little to address the ongoing degradation of the estuary while potentially increasing future tensions around the management of Natura 2000 sites.

2.1.2 A New Approach to Biodiversity Offsetting in the Case of a Large Railway Construction

The design, approval and construction of a new high-speed railway line in southern France provided an opportunity to develop and test new approaches to mitigating and offsetting impacts on biodiversity from infrastructure projects.

The project, entitled *Contournement de Nîmes et Montpellier* (henceforth CNM), included the construction of 80 km of new high-speed track between Nîmes and Montpellier. CNM is run as a public-private partnership. In 2012 construction and maintenance of the railway was delegated to a private consortium of industrial and business corporations named Oc'Via. The duration of this public-private partnership contract is for 25 years (from 2012 to 2037). Construction of the line has been completed in 2017 (Calvet and Quétier 2014).

The railway line crosses two large Natura 2000 sites, one of which holds France's largest population of little bustards (*Tetrax tetrax* L.). This bird became the central concern for this project's biodiversity impact due to the significance of damage to its habitat and conservation status (Calvet and Quétier 2014). The little bustard, which normally inhabits steppes, has become highly dependent on relatively extensive agricultural land uses. Despite some avoidance and reduction measures set up to minimize the project's impact, for example by scheduling construction to minimize disturbance during reproduction, compensation had to be provided for residual impacts. Impacts from the project include direct habitat loss (project footprint) and degradation (from disturbance during construction and operation). Overall, the railway project was projected to impact 1886 ha of natural habitats due to disturbance-related impacts, and destroy 652 ha through construction work. 80% of the impacted area is agricultural land with the rest a mix of Mediterranean woodland and shrubland as well as riverine habitats.

In this context, the need, type and size of biodiversity offsets was assessed using a loss-gain approach based on habitat quality for the little bustard (Quétier et al. 2015). This innovative method, developed by the consultancy firm Biotope, was an alternative to area-based multipliers (as discussed in Sect. 1.2.2). Although loss-gain approaches are likely to enable a more transparent assessment of the feasibility of achieving no net loss (as recommended by BBOP), the method is highly reliant on knowledge of the ecological requirements of the target species as well as data on species presence and habitat features. As a result, it has been criticized as inaccessible to non-specialist audiences, as costly and not always applicable. There is an active debate in France on ensuring that the effort put into designing and sizing biodiversity offsets remains commensurate with the aims of conservation, and that stakeholders are properly trained to apply the right methodological approaches.

Bespoke metrics called compensatory units (CUs) were devised to quantify losses and gains. The main objective of this approach was to highlight how habitat destruction and degradation would generate different levels of loss of habitat for the species, and how different actions would provide varying levels of "gain" (number of CUs) for the targeted species, depending on the initial and final type as well as quality of the habitat that would be conserved, managed or restored as part of the project's offsets. For instance, impacts on a high value habitat will require more CUs to be offset than impacts on a lower quality habitat. Similarly, biodiversity offsets that provide a higher increase in habitat quality will deliver a greater amount of CUs. Through this approach, a total of 3279 CUs was determined as suitable compensation for the impacts of the railway project; most of these units are for little bustards and farmed habitats (95%). On the basis of its operating concession of 25 years, the developer Oc'Via will have to maintain the total of 3279 CUs until 2037. It was estimated that around 1600 ha would be needed to produce the required CUs. This included 500 ha to be acquired and restored as favourable habitats for little bustards, and 1100 to be contracted with willing farmers who agreed to change their management practices to benefit the species. The company set up a mechanism analogous to existing agri-environmental payments under the Common Agricultural Policy, whereby payments to one or several farmer(s) are conditional to changes in their practices that generate CUs (11 different types of agri-environmental measures were proposed to farmers). As of April 2015, offsets included around 1100—1200 ha under contract with 100 farmers covering over 500 plots (average contract is for 11.6 ha), in addition to 512 ha of farmland purchased for conversion into favourable land-cover and leased to farmers for management. Offsets also include 8 ha of woodland (planted), on public land (no purchase) and 136 ha of Mediterranean shrubland purchased and managed for biodiversity.

Using contracts with farmers to achieve offset goals is a somewhat novel approach. Compared to land purchases, contracting with farmers brings many benefits as well as risks. Regarding the benefits, contracts facilitate access to land in a context where land purchases generate significant social conflict especially with farmers. Second, as the design offers greater flexibility, contracts make the offsetting strategy adaptable to environmental or institutional changes (Calvet et al. 2017). They also lower the upfront costs for the developer (while bringing a different kind of financial risk). However, this approach has also limitations.[4] Based on the voluntary nature of contracts from the farmer's perspective, the effectiveness of the offsetting program largely depends on gaining the farmers' participation, yet getting them to sign up is costly. Significant and visible positive effects on species will only be achieved if enough land is under contract at any given time, and for a period of more than two decades. In regard to this point, we note that the second limitation of this scheme is the short duration of contracts, i.e. 5 years in the case of CNM, which makes the offsets vulnerable to changes in the socio-economics of local farming (but this feature is a major part of the attraction of contracting with farmers where landowners were unwilling to sell). Third, the achievement of no net loss depends on the actual implementation of the measures by farmers. In 2015, a specific study has been conducted to assess this (Calvet et al. 2017). The results have revealed a high rate of non-compliance with contracts requirements. Moreover, some required practices were actually already set up on the enrolled plots before signing biodiversity contracts, thus not allowing any gains for offsetting losses of the CNM project. Indeed, as under the well-established Common Agricultural Policy (CAP) of agri-environmental payments, farmers with lower opportunity costs are attracted to the scheme, a fact which may result in windfall effects cancelling out the additional ecological effects of the offsetting program. These results raised concerns about the additionality of this scheme and emphasized the importance of monitoring and sanctions to encourage farmers to meet their commitments. To conclude, while

[4]For more information on the challenges of this contracting approach to offsets, see Calvet et al. (2017).

this new approach to biodiversity offsets may represent an opportunity, important limitations need to be considered to improve the system's effectiveness as a new way to implement biodiversity offsets.

2.2 *Case Study:* Cossure, *France's First Habitat Bank*

CDC Biodiversité was set up in 2008 as a private subsidiary of *Caisse des Dépôts et Consignations* (CDC), a public financial institution. One of its objectives is to provide turnkey solutions to biodiversity offsets for impacts from development projects, including those financed by CDC. Overseen by France's Ministry of the Environment, *CDC Biodiversité* initiated the first pilot habitat bank in 2010, named *Opération Cossure*, in the southeast of France.

In 2008 *CDC Biodiversité* bought an abandoned industrial orchard of 357 ha in the Plaine de Crau to recreate an herbaceous sheep-grazed habitat for steppe birds, of which the most typical are the pin-tailed sandgrouse (*Pterocles alchata*), the Eurasian stone curlew (*Burhinus oedicnemus*), the little bustard (*Tetrax tetrax*) and the calandra lark (*Melanocorypha calandra*). On-site actions included the removal of fruit trees, poplar hedges and irrigation pipes as well as some reshaping of the topography. No new vegetation was planted on the site; instead, natural regeneration from seed stocks already present in the soil and from the adjacent steppes was favoured. The ancient orchard was located within a network of protected areas of steppe forming a nature reserve. Another key objective of this project was to reestablish a continuous grassy cover between the natural reserve areas and this site. The project has also provided an opportunity to conduct some additional experiments in ecological restoration,[5] and to improve knowledge on plants and entomofauna of steppe ecosystems in partnership with local universities. Two sheepfolds were constructed on the site for two herds of sheep brought in to feed on the newly created grasslands. The grazing of 800 ewes is specifically aimed at maintaining a low grass height required for the targeted species of the project. Shepherds are bound by a contract regarding grazing schedules and areas, forbidden animal medications, etc. The natural reserve's managers and agricultural services are also involved in the management and monitoring of the site.

Ecological gains from *Opération Cossure* are calculated on an area basis, namely a total of 357 biodiversity units on 357 ha. After regulatory approval, *CDC Biodiversité* began to sell biodiversity units to developers whose impacts on biodiversity had not been compensated (i.e. ecological debt from past projects) as well as new projects. In order to comply with ecological equivalence requirements, eligible impacts for the purchase of offset units from *Opération Cossure* have to be linked to the same habitats and/or species that are managed by the habitat bank. In addition, impacts have to occur within an informally defined "service area" of 600 km^2. For

[5]These actions are not part of the compensation plan.

each purchase, regulators approve the number of units the developers must buy. CDC *Biodiversité* fixes the price of the units, which are largely determined by the costs of implementation (see below). From an organizational perspective, the management and monitoring of the compensation units is part of *CDC Biodiversité*'s 30-year commitment.

A number of reservations can be expressed in regard to the design of this habitat bank. First, due to the experimental nature of this project, *CDC Biodiversité* has no obligation to continue the management of the unsold units after the end of the initial agreement with the French government in 2016 (which may, however, be renewed). Second, even for the units of the habitat bank that have been sold, there are no binding commitments beyond 30 years of management that raises concerns about the durability of ecological gains. The lack of clear and defined rules in the French regulatory framework related to banking schemes remains an important gap in ensuring that habitat banks provide long term solutions to permanent losses of biodiversity. The site in question may, however, be handed over to a nature conservation agency or NGO (involved in an adjacent natural reserve) for long-term stewardship.

The ecological aspects of the project also raise some interesting questions. The grasslands of *Opération Cossure* were compared to the adjacent natural reserve as a control plot. While the vegetation and its entomofauna have globally recovered since 2008, their interannual variability is still significantly higher than that of the native steppe. Shepherds struggled at times to keep grass growth in check due to particularly high rainfall and insufficient grazing pressure. Although target bird species rapidly colonized the site, their numbers did not consistently grow. External drivers such as grazing in the broader landscape and weather patterns play an essential role in determining the presence and abundance of bird species in the habitat bank. While it is too early to fully assess the conservation outcomes of *Opération Cossure*, the outlook is still positive.

From *Opération Cossure's* total budget of about 12.5 million euros, 44% was spent on land acquisition, 34% on restoration work and 22% on long-term management and monitoring. The price of one biodiversity unit, set by *CDC Biodiversité*, has risen since the first sale of units in 2010 from about 37 k€–41 k€ (Calvet et al. 2015). Projects which have bought units from the habitat bank are logistics platforms as well as one case of *ex post* compensation for accidental damage from a pipeline accident. By 2016 44% of the units had been sold for a total of about 6100 k€ (excluding tax), which is equivalent to a return on investment of 49%. These economic results were worse than expected mainly due to the reduced number of local development projects in the wake of the general economic downturn. In this context, the bank operator had to make some changes in order to boost the sale of biodiversity units. For instance, the bank sold biodiversity units to compensate for impacts on the ocellated lizard (*Timon Lepidus*), a species not initially targeted by the bank. Furthermore, the bank operator tried selling biodiversity units to developers for projects outside the originally agreed 'service area' (Calvet et al. 2015). This illustrates the limitations of the habitat banking to provide units only offsetting impacts on specific and rare species, as discussed in Van Teeffelen et al. (2014). Such a business model is unlikely to be economically viable.

As France's pilot habitat bank, the aim of *Opération Cossure* was to test a new approach to biodiversity offsets inspired by the experiences of the USA mitigation banking. It proved useful in identifying strengths and weaknesses, as well as getting stakeholder feedback on the scheme. *Opération Cossure* provided helpful insights for the inclusion of habitat banking in the 2016 law on biodiversity, although not all weaknesses have been addressed (Vaissière et al. 2017). For more information on this valuable case study, see Calvet et al. (2015) and Dutoit et al. (2015).

2.3 Focus on Marine Biodiversity Offsets

France has the second largest Exclusive Economic Zone in the world. Clearly, a large number of economic activities have impacts on marine biodiversity, including the construction of port infrastructure, dredging, marine aggregate extraction and, increasingly, offshore renewable energy projects. To ensure the successful implementation of marine offsets, as well as the mitigation hierarchy in general, it is necessary to consider legal, ecological and operational specificities of the marine environment.

Contrary to onshore offsetting, which must address complex land tenure issues, marine offset frequently occurs on the Maritime Public Domain[6] (MPD), which is under State jurisdiction. On the one hand, having the State as a single interlocutor facilitates the offsetting process. On the other hand, the status of the MPD (inalienable and imprescriptible, precarious and revocable, all projects subject to an authorization procedure, 'occupation' only temporary) constrains the long-term management of offset measures.

The marine environment also brings challenges to the sizing of offsets. Baseline studies are costly and there are many more knowledge gaps to contend with than on land (e.g. Boehlert and Gill 2010; Carlier 2015).

From an operational perspective, this leads to a restricted cluster of offset actions that generally target a selection of species and habitats such as some threatened species, coral reefs and seagrass ecosystems (Bas et al. 2016).

To date, few marine development projects have actually implemented offset measures (Jacob et al. 2016a). Different types of measures can be identified and described as follows:

- Transplantation measures (e.g. corals, *Cymodocea*, large oysters), which are presented in all regulatory studies as a compensatory measure although they can rather be considered as reduction measures. They are aimed at limiting residual losses, not compensating for them.

---

[6]The MPD, which is composed of the soil and the subsoil of the sea, lies between the upper limit of the shore and the boundary of the territorial sea (12 miles).

- Ecological engineering measures, such as the restoration of coral reefs by larvae collected from colonies after spawning and grown in nurseries before transplantation, or the establishment of artificial reefs to offset the loss of hard substrate. While these have been trialed, their effectiveness is hard to assess.
- Management measures, such as the creation and management of marine protected areas (MPA), are increasingly considered in order to reduce anthropogenic pressures from third parties on degraded ecosystems. Their usage reflects recent international trends. However, this practice should be treated with some caution because of the risk of non-compliance with the additionality criterion (Maron et al. 2015). For example, some projects offset impacts on marine birds by focusing on breeding colonies on land, using similar approaches to those developed for land-based projects.
- Knowledge acquisition, including the funding of monitoring programs (e.g. banding the population of shorebirds), is often implemented in order to improve our understanding of marine ecosystems and the limiting factors for the population recovery of threatened species. Greater insight can increase concerns about the conservation outcomes and the capacity to achieve no net loss.
- Environmental education, such as awareness campaigns on the importance of *Posidonia oceanica* seagrass beds in the Mediterranean, is regularly included in compensation strategies. This also can serve to raise concerns about the conservation outcomes for impacted habitats and species and the capacity to achieve no net loss.

In conclusion, Environmental Impact Assessments highlight in a quasi-systematic way the lack of knowledge on the impacts of marine development projects on marine ecosystems (cf. Vaissière et al. 2014). In a handful of cases, offset measures are proposed because of a misunderstanding of residual impacts and a lack of ecological engineering techniques to offset these impacts. Apart from those offsets actually based on the restoration of degraded ecosystems through ecological engineering measures, current practice concerning marine biodiversity offsets can be fairly criticized. There is scope for marine projects to catch up with recent progress made on land.

3 Conclusions from the Perspective of French Practitioners Regarding an EU-wide No Net Loss and Offset Strategy

Regulation is the key driver of change in the way biodiversity is taken into account in development projects. The reforms of 2007 and 2012 introduced sweeping changes that are only now being widely implemented. Offsets aimed at achieving no net loss for some biodiversity features are now commonly applied in sectors such as transport infrastructure, wind energy and open-cast mining. Several sectoral guidelines have been produced. However, as issues are still identified and managed on a case-by-case basis, the focus given to biodiversity still greatly depends on the specific circumstances of each project (profitability, political support, level of civic opposition, legal precedents, etc.). In general, such considerations determine whether offsets are required and whether a comprehensive 'no net loss' solution is sought and enforced. In this context, housing and urban development projects are clearly falling behind in the use of offsets, as are many of France's overseas territories. While the 2016 law provides additional tools, it does not attempt to overcome the heterogeneity between regions and sectors in the application of the mitigation hierarchy to biodiversity. Of course, time is needed to raise standards, yet good practice is already spreading to achieve the goals spelt out in the law.

In a context of relatively rapid regulatory change, innovative solutions are emerging that make it an exciting time to work in the assessment, mitigation and offsetting of development impacts on biodiversity. The adoption of loss-gain approaches to demonstrate 'no net loss' means that more reliable quantitative methods are being used to determine the presence and abundance of species or variations in habitat quality. At the same time the cost of offsetting is increasing as stakeholders better understand what is at stake and oversight of implementation improves. As biodiversity offsetting is more aggressively applied, businesses and investors are getting better at anticipating the associated risks and costs of development. Indeed, some local governments (mainly in urban areas with strong development pressure) are now integrating application of the mitigation hierarchy to both project and plan level in order to avoid deadlocks. Such projects are being set-up as part of (or outside) the national habitat banking "experiment", somewhat akin to Germany's "pools" system. This bottom-up dynamic is encouraging, even if it means that some solutions will perform poorly and considerable blind spots remain.

In conclusion, the past decade has seen rapid change in both regulations and the practical implementation of the mitigation hierarchy in France. Although different sectors and regions are moving forward at different speeds, the overall direction is towards a healthier balance of development and biodiversity. In this context, biodiversity offsets are seen as a useful tool, with many technical and organizational solutions being developed and tested. It is too early to tell if these will perform as expected. Success very much depends on the continued engagement of citizens to foster better long-term outcomes for biodiversity in the context of development.

References

Aulert C, Provost P, Bessineton C, Dutilleul C (2009) Les mesures compensatoires et d'accompagnement Port 2000: retour d'expériences. Ingénierie N° spécial: 55–72

Barnaud G, Coïc B (2010) Mesures compensatoires et correctives liées à la destruction des zones humides: revue bibliographique et analyse critique des méthodes. Convention ONEMA–MNHN

Bas A, Jacob C, Hay J, Pioch S, Thorin S (2016) Improving marine biodiversity offsetting: a proposed methodology for better assessing losses and gains. J Environ Manag 175:46–59

Bigard C, Pioch S, Thompson JD (2017) The inclusion of biodiversity in environmental impact assessment: policy-related progress limited by gaps and semantic confusion. J Environ Manag 200:35–45

Boehlert GW, Gill AB (2010) Environmental and ecological effects of ocean renewable energy development. Oceanography 23(2):68–81
Bull JW, Brauneder K, Darbi M, van Teeffelen AJA, Quétier F, Brooks S, Dunnett S, Strange N (2018) Data transparency regarding the implementation of European 'no net loss' biodiversity policies. Biol Conserv 218: 64-72. https://doi.org/10.1016/j.biocon.2017.12.002
Briggs S, Hudson MD (2013) Determination of significance in ecological impact assessment: past change, current practice and future improvements. Environ Impact Assess Rev 38:16–25
Bristow G, Cowell R, Munday M (2012) Windfalls for whom? The evolving notion of 'community' in community benefit provisions from wind farms. Geoforum 43(6):1108–1120
Calvet C, Quétier F (2014) La prise en compte de la biodiversité dans le contournement ferroviaire Nîmes-Montpellier: Qui fait quoi? Revue Espaces Naturels 45:23–25
Calvet C, Levrel H, Napoléone C, Dutoit T (2015) La réserve d'actifs naturels. Une nouvelle forme d'organisation pour la préservation de la biodiversité en France? In: Levrel H, Frascaria-Lacoste N, Hay J, Martin GJ, Pioch S (eds) Restaurer la nature pour atténuer les impacts du développement – Analyse des mesures compensatoires pour la biodiversité. QUAE, Versailles, pp 199–209
Calvet C, Le Coent P, Napoléone C, Quétier F (2017) Challenges of achieving biodiversity offset outcomes through agri-environmental schemes: evidence from an empirical study in Southern France. DR n°2017-05, Working Paper, UMR LAMETA, INRA, Montpellier
Carlier A (2015) Les actions de restauration des écosystèmes benthiques côtiers. In: Levrel H, Frascaria-Lacoste N, Hay J, Martin GJ, Pioch S (eds) Restaurer la nature pour atténuer les impacts du développement – Analyse des mesures compensatoires pour la biodiversité. QUAE, Versailles, pp 199–209
Camesella C, Caritg J-P, Chédot C, Maillard M (2013) 32,000 emplois sur le complexe industrialo-portuaire du havre, AVAL 132
DCGSD (2011) The new environmental liability mechanism: equivalence methods for repairs "in kind". Le point sur 96
De Billy V, Tournebize J, Barnaud G, Benoît M, Birgand F, Garnier J, Lesaffre B, Lévêque C, de Marsily G, Muller S (2015) Compenser la destruction de zones humides. Retours d'expérience sur les méthodes et réflexions inspirées par le projet d'aéroport de Notre-Dame-des-Landes (France). Natures Sciences Sociétés 23:27–41
De Marsily G, Barnaud G, Benoit M, De Billy V, Birgand F, Garnier J, Le Saffre B, Leveque C, Muller S, Musy A, Tournebize J, Zimmer D (2013) Rapport du collège d'experts scientifiques relatif à l'évaluation de la méthode de compensation des incidences sur les zones humides. In: Projet de réalisation de la plateforme aéroportuaire, du programme viaire d'accompagnement et de la desserte routière de l'aéroport de Notre-Dame-Des-Landes
DREAL PACA (2009) Les mesures compensatoires pour la biodiversité. Principes et mises en œuvre en région PACA
Dutoit T, Calvet C, Jaunatre R, Alignan JF, Wolff A, Sauguet F, Bulot A, Buisson E, Debras JF, Provost E, Napoléone C (2015) Première expérimentation de compensation par l'offre : bilan et perspective. Sciences Eaux & Territoires 16, L'ingénierie écologique au service de l'aménagement du territoire
Enviroscop, CERE, SOGREAH, In Vivo (2010) Analyse de mesures compensatoires aux atteintes au patrimoine naturel. Recueil et analyse de cas. Rapport rédigé pour le Ministère de l'Écologie, Paris
Gaucherand S, Laroche A, Quétier F (2013) De l'évaluation des écosystèmes à la mise en oeuvre des mesures environnementales : Bilan des études environnementales réalisées en station de ski entre 2000 et 2012. Document de travail IRSTEA & Biotope
Gayet G, Baptist F, Baraille L, Caessteker P, Clément JC, Gaillard J, Gaucherand S, Isselin-Nondedeu F, Poinsot C, Quétier F, Touroult J, Barnaud G (2016) Guide de la méthode nationale d'évaluation des fonctions des zones humides – version 1.0. ONEMA, collection Guides et protocoles
Gobert J (2010) Ethique environnementale, remédiation écologique et compensations territoriales. VertigO – La revue en sciences de l'environnement 10(1)
Groupe de travail sur le préjudice écologique (2013) Pour la réparation du préjudice écologique. Available via http://www.ladocumentationfrancaise.fr/var/storage/rapports-publics/134000619.pdf. Accessed 24 Aug 2015

Jacob C, Pioch S (2014) Protocoles expérimentaux d'aide aux services instructeurs dans l'analyse de dossiers compensatoires. ONEMA (ed)

Jacob C, Pioch S, Thorin S (2016a) The effectiveness of the mitigation hierarchy in environmental impact studies on marine ecosystems: a case study in France. Environ Impact Assess Rev 60:83–98

Jacob C, Vaissière AC, Bas A, Calvet C (2016b) Investigating the inclusion of ecosystem services in biodiversity offsetting. Ecosyst Serv 21:92–102

Kermagoret C, Levrel H, Carlier A (2014) The impact and compensation of offshore wind farm development: analysing the institutional discourse from a French case study. Scott Geogr J 130(3):188–206

Le club des juristes (2012) Mieux réparer le dommage environnemental. Available via http://www.leclubdesjuristes.com/wp-content/uploads/2014/03/Rapport_Commission_Environnement-final.pdf. Accessed 24 Aug 2015

Levrel H, Frascaria-Lacoste N, Hay J, Martin GJ, Pioch S (eds) (2015) Restaurer la nature pour atténuer les impacts du développement – Analyse des mesures compensatoires pour la biodiversité. QUAE, Versailles

Maron M, Gordon A, Mackey BG, Possingham HP, Watson JEM (2015) Conservation: stop misuse of biodiversity offsets. Nat Comments 523:401–403

Martin GJ (2014) L'entrée de la réparation du préjudice écologique dans le code civil: les projets en droit français. Revue Générale des Assurances et des Responsabilités 4:15063

MEDDE (2013) Lignes directrices nationales sur la séquence éviter, réduire et compenser les impacts sur les milieux naturels

MEDTL (2012) Doctrine relative à la séquence éviter, réduire et compenser les impacts sur le milieu naturel

Moreno-Mateos D, Power ME, Comín FA, Yockteng R (2012) Structural and functional loss in restored wetland ecosystems. PLoS Biol 10(1):e1001247

Mudgal S, Chenot B, Salès K et al (2013) Implementation challenges and obstacles of the Environmental Liability Directive (ELD) Final Report. Available at http://ec.europa.eu/environment/archives/liability/eld/eldimplement/pdf/ELD%20implementation_Final%20report.pdf. Accessed 24 Aug 2015

Neyret L, Martin GJ (eds) (2012) Nomenclature des préjudices environnementaux. LGDJ, Paris

Quétier F, Lavorel S (2011) Assessing ecological equivalence in biodiversity offset schemes: key issues and solutions. Biol Conserv 144:2991–2999

Quétier F, Regnery B, Levrel H (2014) No net loss of biodiversity or paper offsets? A critical review of the French no net loss policy. Environ Sci Pol 38:120–131

Quétier F, Moura C, Menut T, Boulnois R, Rufray X (2015) La compensation écologique fonctionnelle: innover pour mieux traiter les impacts résiduels des projets d'aménagements sur la biodiversité. Sci Eaux Territ 2:24–29

Regnery B, Quétier F, Cozannet N, Gaucherand S, Laroche A, Burylo M, Couvet D, Kerbiriou C (2013) Mesures compensatoires pour la biodiversité: comment améliorer les dossiers environnementaux et la gouvernance ? Sciences, Eaux et Territoires Numéro hors-série

Scemama P, Levrel H (2016) Using habitat equivalency analysis to assess the cost-effectiveness of restoration outcomes in four institutional contexts. Environ Manag 57:109–122

UICN France (2011) La compensation écologique: État des lieux et recommandations. Paris

Vaissière AC, Bierry A, Quétier F (2016) Mieux compenser les impacts sur les zones humides: modélisation de différentes approches dans la région de Grenoble. Sci Eaux Territ 22:14–19

Vaissière AC, Quétier F, Levrel H (2017) Le nouveau dispositif des sites naturels de compensation: est-ce trop tôt ? In: Cans C, Cizel O (dir.) La loi biodiversité. Ce qui change en pratique. Avec les témoignages de juristes, chercheurs, avocats, associations. Éditions Législatives, pp 126–135

Vaissière AC, Levrel H, Carlier A, Pioch S (2014) Biodiversity offsets for offshore wind farms projects: the current situation in Europe. Mar Policy 48:172–183

Van Teeffelen AJ, Opdam P, Wätzold F, Hartig F, Johst K, Drechsler M, Vos CC, Wissel S, Quétier F (2014) Ecological and economic conditions and associated institutional challenges for conservation banking in dynamic landscapes. Landsc Urban Plan 130:64–72

Walker B, Wiersma B, Bailey E (2014) Community benefits, framing and the social acceptance of offshore wind farms: an experimental study in England. Energy Res Soc Sci 3:46–54

Germany

Wolfgang Wende, Juliane Albrecht, Marianne Darbi, Matthias Herbert, Anja May, Jochen Schumacher, and Martin Szaramowicz

The Federal Republic of Germany lies at the heart of Europe. It is encircled by nine other countries: Denmark to the north, the Netherlands, Belgium, Luxembourg and France to the west, Switzerland and Austria to the south and the Czech Republic and Poland to the east. Germany's natural boundaries are the Baltic Sea to the north as well as the North Sea. Its population of 82 million citizens inhabit a landmass of

W. Wende (✉)
Leibniz Institute of Ecological Urban and Regional Development and TU Dresden, Dresden, Germany

National University of Singapore, Singapore, Singapore
e-mail: w.wende@ioer.de

J. Albrecht
Leibniz Institute of Ecological Urban and Regional Development and TU Dresden, Dresden, Germany
e-mail: j.albrecht@ioer.de

M. Darbi
Helmholtz Centre for Environmental Research, Leipzig, Germany
e-mail: marianne.darbi@ufz.de

M. Herbert
Federal Agency for Nature Conservation, Leipzig, Germany
e-mail: matthias.herbert@bfn.de

A. May
Berlin, Germany
e-mail: anjamay@gmx.de

J. Schumacher
Tübingen, Germany
e-mail: NuR@springer.com

M. Szaramowicz
Compensation Agency Brandenburg, Brandenburg an der Havel, Germany
e-mail: martin.szaramowicz@flaechenagentur.de

W. Wende et al. (eds.), *Biodiversity Offsets*,
https://doi.org/10.1007/978-3-319-72581-9_8

approximately 357,000 km^2 (Statistical Agencies of the *Bund* and the *Länder* 2016). Germany is a federation made up of 16 largely autonomous federal states: Bavaria, Brandenburg, Saxony Anhalt, Saxony, Thuringia, Mecklenburg-West Pomerania, Schleswig-Holstein, Lower Saxony, Hesse, Rhineland Palatinate, Baden-Württemberg, Saarland, Bavaria and North Rhine-Westphalia as well as the city states of Berlin, Hamburg and Bremen.

Germany is a densely populated country (around 230 per km^2). Its natural landscapes have been manipulated by man from the earliest times (Anhuf et al. 2003, p. 88 ff.). Today we can say that the entire territory of Germany has been transformed into a network of artificial landscapes and semi natural habitats (Siebert and Steingrube 2000 p. 40). In particular, high intensity agriculture as well as the expansion of settlements and infrastructure over the past decades have damaged and indeed in some cases destroyed valuable habitats. One third of Germany's species of flora and fauna have been placed on an official "red list", which means they are considered to be endangered in the country.

Diverse instruments are employed in Germany to preserve landscapes and valuable habitats, as well as to protect their associated species. These are legally underpinned by the *Bundesnaturschutzgesetz* (Federal Act for the Protection of Nature) and additionally by the nature protection laws of the individual states (*Länder*). Typical instruments are the designation of national and European protected areas, the protection of biotopes wherever they occur, the biotope network as well as species protection law. Within the framework of the discussion on No Net Loss of biodiversity, the so-called Impact Mitigation Regulation (IMR) (*Eingriffsregelung*) according to §§ 13 ff. BNatSchG (Federal Nature Conservation Act) plays a key role (see Albrecht et al. 2014; Darbi et al. 2010; Wende et al. 2005).

IMR is a comprehensive instrument to mitigate and offset impacts on *all* forms of nature and landscape, not just those specifically designated as protected sites but also unprotected but still valuable nature assets and the "normal" landscape. The aim is to safeguard the functional capacities and services of nature as well as the integrity of the country's landscapes. Therefore, the Impact Mitigation Regulation adopts a different statutory framework than required, for example, to secure network coherence within the FFH habitats assessment. Whereas models from other member states described in this book have not been fully realised in all cases, the German system of biodiversity offsetting has already been strictly implemented.

1 Theory in Germany

1.1 Legal Background in Germany

1.1.1 Impact Mitigation Regulation Under Conservation Law

The Impact Mitigation Regulation (IMR) was introduced in Germany in 1976 under the Federal Nature Conservation Act (BNatSchG 1976). It has been subsequently revised on several occasions. From the outset, IMR has been one of the core elements of the country's conservation law. In the current version of the BNatSchG from 29 July 2009, the respective stipulations on IMR are contained in §§ 13–18 (BNatSchG 2009). As a result of the new competing legislative competence of the federal government in conservation law (Art. 74 Para. 1 No. 29 Grundgesetz), which was implemented in the context of the federalism reform of 2006 (GG 2006), these are comprehensive regulations possessing full legal force. State-level regulations implementing the Impact Mitigation Regulation have been adapted to the new competences (see Table 1). Complementary state-level regulations as well as those which deviate from federal regulations in a permissible manner (cf. Art. 72 Para. 3 No. 2 Grundgesetz) do, however, remain in force.

Under § 13 BNatSchG, Impact Mitigation Regulation is a general principle of conservation law. § 13 BNatSchG establishes the responsibilities of parties who cause negative impacts on nature and landscapes. The primary aim is that such harm to nature and landscapes be avoided as far as possible. If in-kind compensation or offsetting measures cannot be implemented, then the party responsible for the damage must pay financial compensation (cf. Schumacher/Fischer-Hüftle, § 13 BNatSchG Rdnr. 1). This statutory regulation therefore adopts the basic principle of the "polluter pays" principle as well as the prioritisation of obligations on parties encroaching on nature. Specifically, in-kind compensation and financial compensation are not viewed as equivalent (cf. Schumacher/Fischer-Hüftle, § 13 BNatSchG Rdnr. 4).

Accordingly, the first priority for any party implementing a development project is to follow the mitigation hierarchy (see also chapters "Introduction of a European Strategy on No Net Loss of Biodiversity" and "Principles of the Business and Biodiversity Offsets Programme" with the offsetting principles) to avoid significant negative impacts upon nature or the landscape. Unavoidable significant negative impacts are to be compensated by in-kind and on-site compensation or in-kind/out-of-kind and off-site offsetting measures. If these are not feasible, then offset payments are foreseen. While the states may not deviate from this general principle (Art. 72 Para. 3 No. 2 Grundgesetz), they have the option of complementing, revising or deviating from specific national regulations that specifically accommodate such deviations. Examples of state-level provisions are so-called eco-account directives (defining some forms of habitat banks) and compensation directives. These deal with the legal preconditions for compensation site cadastres (and/or property registers) and eco-accounts as well as methods to evaluate negative impacts and their compensation. Furthermore, nature conservation laws of the *Länder* serve to determine the competent authorities and the procedure of Impact Mitigation Regulation.

Existence of an Impact as a Precondition

Under § 14, Para. 1 BNatSchG, impacts upon nature or the landscape are defined as activities which change the form or use of land areas or of aquifers connected with the biospheric soil stratum in a manner that considerably undermines the efficiency of the balance of nature or the landscape scenery.

Table 1 List of nature conservation laws of Germany's constituent states

Federal state	Nature conservation law	Regulations on IMR
Baden-Württemberg	Baden-Württemberg Nature Conservation Law of 23 June 2015 (NatSchG Ba-Wü)	§§ 14–21
Bavaria	Bavarian Nature Conservation Law of 23 February 2011 (BayNatSchG)	Art. 6–11
Brandenburg	Brandenburg Nature Conservation Implementation Law of 21 January 2013 (BbgNatSchAG)	§§ 6, 7
Berlin	Berlin Nature Conservation Law of 29 May 2013 (NatSchG Bln)	§§ 16–19
Bremen	Bremen Nature Conservation Law of 27 April 2010 (BremNatG)	§§ 8, 9
Hamburg	Hamburg Implementation Law of the Federal Nature Conservation Act of 11 May 2010 (HmbBNatSchAG)	§§ 6–8
Hesse	Hessian Implementation Law of the Federal Nature Conservation Act of 20 December 2010 (HAGBNatSchG)	§§ 7–11
Lower Saxony	Lower Saxon Implementation Law of the Federal Nature Conservation Act of 19 February 2010 (NAGBNatSchG)	§§ 5–7
Mecklenburg-West Pomerania	Mecklenburg-West Pomeranian Implementation Law of the Federal Nature Conservation Act of 23 February 2010 (NatSchAG M-V)	§§ 12, 13
North Rhine-Westphalia	Nature Conservation Law of the Land North Rhine-Westphalia of 15 November 2016 (LNatSchG NRW)	§§ 30–34
Rhineland Palatinate	Nature Conservation Law of the Land Rhineland Palatinate of 6 October 2015 (LNatSchG R-Pf)	§§ 6–10
Saarland	Saarland Nature Conservation Law of 5 April 2006 (SNG)	§§ 27–30
Saxony	Saxon Nature Conservation Law of 6 June 2013, SächsNatSchG	§§ 9–12
Saxony-Anhalt	Nature Conservation Law of the Land Saxony-Anhalt of 10 December 2010 (NatSchG LSA)	§§ 6–14
Schleswig-Holstein	Nature Conservation Law of the Land Schleswig-Holstein of 24 February 2010 (LNatSchG S-H)	§§ 8–11a
Thuringia	Thuringian Nature Conservation Law of 30 August 2006 (ThürNatG)	§§ 6–10

The Impact Mitigation Regulation, therefore, is not restricted to defined forms of human intervention but basically applies to all activities (with the exception of agriculture and forestry) that have a significant negative impact on nature and the landscape. The concept of impact consists of two components: firstly, consideration of the cause of the impact (*Ursache*), and, secondly, establishment of the potential for negative effect (*mögliche Wirkung*) (Koch 2010, p. 133). Both of these components have to be determined before the IMR can be applied (cf. Schumacher/Fischer-Hüftle, § 14 BNatSchG Rdnr. 2).

The form of land areas according to § 14, Para. 1 BNatSchG means both the geomorphological phenomenology and the stock of vegetation in the areas concerned (Lau 2011a, p. 682). Typical changes in the shape of land include those caused by residential/commercial developments or new infrastructure. "Use" refers to any utilization of an area for a certain purpose (Lau 2011a, p. 682). A change in use exists if "an existing use is replaced by another" (Koch 2010, p. 134). Examples of such change are the construction of structures on existing open spaces or the transformation of grassland into farmland.

Under § 7 Para. 1 No. 2 BNatSchG, the various components of the natural balance are the soil, water, air, the climate, plants and animals as well as the interaction between these. A negative impact on the landscape scenery is said to exist if changes are such as to be perceived as detrimental by a typical observer sensitive to the beauty of a naturally developed landscape (BVerwG, Ruling of 27.9.1990; OVG Münster, Ruling of 10.02.1998).

§ 14, Para. 2 of the BNatSchG lays out the so-called "agricultural privilege", which exempts the use of soil for agricultural, silvicultural and fishing purposes from Impact Mitigation Regulation provided that these correspond with good professional practice according to § 5 Paras. 2–4 BNatSchG. This "agricultural clause" depends (both substantively and systematically) on the definition of "intervention" laid out in § 14 Para. 1 BNatSchG. Hence, an activity is *not* considered to be an intervention if the conditions specified in the agricultural clause are met (Schumacher/Fischer-Hüftle, § 14 BNatSchG Rdnr. 57). Seen from the nature conservation and a more technical point of view this agricultural clause is one of the gaps. However, it was a political decision to design the law as such in the initializing phases of this new legislation in 1976 and during later amendments.

Mitigation Hierarchy

The IMR follows a strict mitigation hierarchy (see Herbert 2015; Wende et al. 2005; Schumacher/Fischer-Hüftle, § 15 BNatSchG Rdnr. 5). The approval of projects depends on compliance with certain requirements and duties, which are variously prioritised (Koch 2010, p. 139). This so-called "cascade of legal sequence" or "mitigation hierarchy" encompasses:

- Requirement to avoid negative impacts where possible (§ 15, Para. 1 BNatSchG);
- The duty of compensation or offsetting for unavoidable negative impacts (§ 15, Para. 2, Clause 1 BNatSchG);
- The requirement of a balanced consideration of interests between unavoidable negative impacts upon nature and the landscape, on the one hand, and interests connected to implementation of the impacting measure, on the other (§ 15, Para. 5 BNatSchG);
- The duty, in the case of project approval after a balanced consideration of interests, to render monetary compensation for unavoidable damages caused that are not offsetable (§ 15, Para. 6 BNatSchG).

Avoidance

Under § 15, Sect. 1, Clause 1 BNatSchG, the party causing the impact has a duty to refrain from any avoidable negative impact upon nature and the landscape. An impact is to be considered avoidable if an acceptable alternative exists that could achieve the purpose of the project at the same location but with less or indeed no impact on nature and the landscape (§ 15, Para. 1, Clause 2 BNatSchG). That means that the duty of avoidance of an impact refers primarily to the type and scope of the project, and is less concerned as to whether the project is fundamentally admissible or not (Weiland and Wohlleber-Feller 2007, p. 277). The duty of avoidance mandates that the party carrying out the project must strive to minimise environmental impacts during the planning and implementation phases (Scheidler 2010, p. 137). If negative impacts cannot be avoided, this fact must be demonstrated as per § 15, Para. 1, Clause 3 BNatSchG.

The respective definition of impact *avoidance* measures and impact *compensation* measures can be deduced from the legal wording. § 15 Para. 1 BNatSchG does not state that adverse effects on nature and landscapes should be "avoided" but rather than the intervening party should "refrain" from causing avoidable harm. This primary obligation to refrain from causing avoidable adverse effects is certainly the best approach to preserving the status quo of nature and landscapes. Compensation measures and offsets are seen as one step lower in the mitigation hierarchy, as these presuppose some negative impact that must be neutralized. Full or partial avoidance of harm means that the construction and operation of some planned development results in no (or merely a low intensity) adverse impact (cf. Schumacher/Fischer-Hüftle, § 15 BNatSchG Rdnr. 26).

Compensation or Offsetting of Unavoidable Negative Impacts

The party implementing a project must compensate (*ausgleichen*) or offset (*ersetzen*) any unavoidable negative impacts by conservation and/or landscape care measures (§ 15, Para. 2, Clause 1 BNatSchG). Whereas (restoration) compensation means in-kind and on site; offset (replacement) means in-kind/out-of kind and offsite.

An impact can be considered *compensated* if and when the negatively affected functions of the balance of nature have been restored in-kind, and the quality of the landscape has been appropriately restored or newly established (§ 15, Para. 2, Clause 2 BNatSchG). Compensation only applies to land areas which in fact require restoration, and which can be effectively upgraded. Moreover, a functionally-specific spatial connection between the impact and the compensation must exist (on-site). A full-coverage minimum level of protection of nature and the landscape is often best achieved if the compensation is primarily implemented at the point of the impact ("on-site, in-kind compensation", also referred to as "restoration compensation"). An impact can be considered *offset* if and when the negatively affected functions of the natural balance have been restored to an equivalent value in the affected area, and the landscape appearance has been appropriately re-designed in a manner consistent with the landscape (§ 15, Para. 2, Clause 3 BNatSchG). Offsetting is also referred to as "replacement compensation".

In contrast to previous versions of the BNatschG which favoured restoration compensation over offsets, the sequence of decisions governing compensation and offsetting have been equivalent since the amendment of 2009. The abandonment of compensation as the favoured approach to impact mitigation does not, however, mean that intervening parties have a free choice between compensation or offsetting on the basis of, for example, financial gain or loss, without considering the goals of Impact Mitigation Regulation. The intended aim of the legal revision was certainly not the standard adoption of offset measures while compensation is relegated to the waste bin (cf. Schumacher/Fischer-Hüftle, § 15 BNatSchG Rdnr. 60). Rather, the responsible authority is obliged to make a decision that is substantively the best and most proportionate (Koch 2010, p. 146). In some cases, therefore, replacement/offset measures *may* take priority if these generate a greater overall benefit for nature and the landscape.

Restoration compensation and offsetting must be carried out "within an appropriate period" (cf. § 15, Para. 5 BNatSchG) to the intervention, preferably simultaneously. Since this is not always feasible in practice, prior compensation and offsetting measures undertaken with regard to anticipated interventions can be booked in so-called "eco-accounts" and "land pools" (i.e. forms of habitat banks). These can subsequently be "withdrawn" as needed (cf. § 16, Para. 2, BNatSchG). The motivation for this rule is, first, to improve the interaction between compensation and offsetting measures, and, second, to remedy the lack of suitable land areas for development, particularly in densely populated areas and in the case of major projects (Herbert and Mayer 2007, p. 17 ff.).

Balanced Consideration of Interests

After investigating how the expected project-related negative impact on nature and the landscape can be reduced (through avoidance, minimisation, compensation, offsetting), the next step is to weigh up the (remaining) potential negative impacts against the interests and potential benefits of the project (§ 15, Para. 5 BNatSchG; Koch 2010, p. 151). Accordingly, an impacting project may not be authorized or implemented if the impacts cannot be mitigated, compensated or offset within a reasonable time period, and if the interests of nature and the landscape are found to take priority over other interests in accordance with a balanced consideration of all demands on nature and the landscape. The responsible authority is obliged to present evidence and bears the burden of proof that priority has been given to the interests of nature and the landscape (Lau 2011b, p. 768). Some may say this might result in projects with otherwise significant impacts now being allowed, and this weakens nature protection. However, in Germany, there is no empirical evidence for this.

The balanced consideration of interests has a corrective function in the scope of IMR by forcing the project developer to provide a sound justification for the necessity of a project (Jessel 2003, p. 121). Originally, any consideration of "offsetting measures" was carried out after the "balanced consideration of interests". Under the BNatSchG amendment of 4 April 2002 (BNatSchG 2002), consideration of "offsetting measures" has—for reasons of practicality—been moved to the planning stage before the "balanced consideration of interests". The version in force before 2002 had the advantage of encouraging "not just 'something or other', but instead, really ambitious and extensive compensation measures" to be carried out (Jessel 2003, p. 121). However, the fact that the offsetting step was moved to a planning stage before the balanced consideration of interests can also be seen as emphasising the priority of natural compensation over offset payments. This also corresponds with the equalisation of compensation and offsetting within the mitigation hierarchy as laid out in § 15, Para. 2, Clause 1 BNatSchG in the version from 2009.

Offset Payments
Under § 15, Para. 6, Clause 1 BNatSchG, if a project that will result in unavoidable impacts that cannot be compensated or offset during an acceptable period of time is nonetheless approved or implemented after consideration of the interests of nature and the landscape as well as the interests of the project implementation, then the intervening party must render an offset payment. The duty to render an offset payment is subsidiary to the compensation and offsetting measures, and hence the "last resort". Under § 15, Para. 6, Clause 1 and 2 BNatSchG, the level of offset payment is calculated according to the average cost of non-implemented compensation and offsetting measures (i.e. according to the theoretical restoration costs; which is only applicable for those habitats that actually can be restored) or according to the duration and severity of the impact. The offset payment is earmarked for measures of conservation and landscape care, preferably in the affected natural area (§ 15, Para. 6, Clause 1 BNatSchG). Specifically, it should be employed for the additional upgrading of nature and the landscape, and not to fund other government conservationist requirements (Koch 2010, p. 159).

The details of compensation for impacts have hitherto been regulated by ordinances of the states. This has resulted in a somewhat confusing legal situation due to the multiplicity of legal sources and some divergences in specific provisions (Koch 2010, p. 166). Based on § 15, Para. 7 BNatSchG, the federal government submitted a draft ordinance on 5 November 2012 regarding compensation for impacts on nature and the landscape, called the Federal Compensation Ordinance (BKompV 2013). The aim is to unify standards and procedures for handling impacts, and thus to make implementation more effective throughout the country. In particular, the ordinance is designed to regulate the details of the content, type and scope of compensation and offsetting measures as well as the level of offset payments and the procedure for imposing them. However, it is currently unclear whether and when the compensation ordinance will come into force.

1.1.2 Impact Mitigation Regulation Under Building Law

Since the introduction of Impact Mitigation Regulation through the BNatSchG (originally promulgated in 1976), there have been points of confusion and disagreement regarding how Impact Mitigation Regulation relates to building law (cf. Federal Building Code—BauGB 1960). These problems are rooted in the fact that, as most interventions in nature and the landscape are building projects, two approval regimes are involved: IMR for issues of nature conservation and the planning approval required by building law (Wagner 2007, p. 114). Here one particular problem proved to be the integration of Impact Mitigation Regulation into the standard development planning procedure (*Bebauungsplanverfahren*) and the relationship to subsequent planning approval. Thus, the question arose whether the impact assessment should take place during the final stage of planning approval or whether it should be realised earlier, when the initial urban land-use plan (*Bauleitplan*) is presented (Dolde 1993; Gaentzsch 1990). The law decided that, rather than a double impact assessment for the land-use plan as well as for the later approval process, only one investigation of the impact on nature is required (IWG 1993). This was realised by the insertion of §§ 8a–8d BNatSchG, termed a "building law compromise", which determined that an independent impact assessment is only required for the approval of developments in so-called "undesignated outlying areas" as specified by § 35 BauGB. No impact assessment was required for most planning approvals, as such assessment had already taken place at the preliminary planning stage.

Despite this clear demarcation, the Impact Mitigation Regulation continued to prove problematic in the real world. In particular, it was unclear to which extent IMR should be considered in the deliberations around the initial land-use plan, and in which specific ways it should be reflected. To resolve this issue, lawmakers further refined the building law compromise (BauROG 1997) by basically removing the IMR as part of the urban land-use plan from the BNatSchG (annulment of §§ 8b and 8c, revised version § 8a) and inserting it into the BauGB (§§ 1a Para. 2 No. 2 and Paras. 3; 5 Para. 2a; 9 Para. 1a; 135a–135c; 200a). The relationship of IMR for nature protection and IMR in regard to urban development was regulated by the newly revised § 8a BNatSchG as follows: The material core of regulation remains a matter of the Federal Nature Conservation Act, whereas the legal consequences of an impact and the various types of mitigation and compensation measures to be implemented are directly regulated by the Federal Building Code. The end result has been to divide the previously unified Impact Mitigation Regulation into one concerned with issues of nature protection and one concerned with the logic of urban planning, i.e. an IMR rooted in building law. The contents of § 8a BNatSchG were adopted into § 21 BNatSchG as part of the BNatSchG revision of 2002, and are now regulated by § 18 BNatSchG 2009.

According to § 1a Para. 3 Clause 1 BauGB, the concerns of nature protection and care of the landscape as well as the avoidance and compensation of projected negative impacts on the landscape as well as the services and capacity of the nature balance must be considered when drawing up an urban land-use plan as specified by § 1 Para. 7 BauGB. If major impacts on the balance of nature are forecast for certain projects according to the planning presentation or stipulations for plan implementation, then these impacts must be compensated by the simultaneous presentation or designation of compensatory areas in the preparatory land-use plan or building plan or other suitable measures implemented within areas provided by the municipality (§ 1a Para. 3 Clause 2 and Clause 4 BNatSchG, cf. Schrödter, § 1a BauGB Rdnr. 44 ff.). The practical realisation of compensation can be agreed by an urban development contract according to § 11 BauGB.

The governing criteria for the avoidance and compensation of a projected negative impact on nature and the landscape within land-use planning are the principles of a fair balancing of all relevant public and private concerns. However, it can happen that compensation measures are discarded during this weighing up of all interests (Meyhöfer 2000). In contrast to Impact Mitigation Regulation under §§

13 ff. BNatSchG, land-use planning is not required to ensure full compensation (Götze and Müller 2008). Hence, the complete compensation of an intervention, which is the express aim of Impact Mitigation Regulation, may not be realised (Balla et al. 2000). No direct spatial connection is demanded between the site of impact and the site of compensation, insofar as this is compatible with well-ordered urban development, the aims of spatial planning as well as those of nature protection and the conservation of the landscape (§ 1 Para. 3 Clause 3, § 200a Clause 2 BauGB).

1.2 Methodological Background in Germany

The following methodological background to impact mitigation focuses on the calculation of metrics as well as measures to conserve landscape quality. In Germany practical application of the IMR/*Eingriffsregelung* is often interlinked with landscape planning and plans (see Makala 2016; von Haaren et al. 2008 or Tucker et al. 2013, p. 163). Landscape plans provide a comprehensive and integrative solution to the questions of where, how and in which priority offsets can actually be implemented at the local level, and in which green infrastructure network they will prove most effective. Generally, strategies of mitigation and offsetting are clearly necessary to ensure good no net loss results.

1.2.1 Metrics and Typical Offset Measures in Germany

Metrics, as defined by the glossary of the Working Group on No Net Loss, are a "set of unitary measurements of biodiversity lost, gained or exchanged. These are used in order to compare losses at the damaged site and gains at the compensation site and provide decision support to establish equivalence" (Dickie et al. 2013). In Germany, metrics are used to calculate impacts and offsets for the following subjects of protection: Habitats/biotopes, species, soil, water, air and climate (functions), as well as landscapes in terms of their aesthetic natural value and in some cases their recreational value (see Rayment et al. 2014).

Metrics vary from very basic measures such as area analysis, to sophisticated quantitative indices of multiple biodiversity components, which may be variously weighted. A wide variety of metrics are employed in Germany, reflecting the different approaches of the federal states. However, these can be roughly categorized into four basic types (Bruns 2007; Darbi and Tausch 2010 or Köppel et al. 2004):

- Compensation area coefficients/ratios for biotope types and functions (i.e. a habitat area approach).
- Biotope valuation procedures, including those based solely on the ecological value and area of biotopes as well as those that reflect adjustments to varying degrees, based on the condition of the biotopes (i.e. habitat area × value/conditon).

- The calculation of replacement costs, i.e. an estimate of the cost of restoring the impacted area.
- Qualitative reasoning (verbal argument-based methods).

The metrics applied in some federal states integrate several of these approaches (see Rayment et al. 2014). Most often one finds biotope valuation procedures combining an ecological valuation of a habitat with its area. The following Table 2 from the Bavarian guidelines for mitigation regulation shows which restoration compensation and biodiversity offset measures are typical for mitigation regulation in Germany.

1.2.2 Methodology of Habitat Banking/Compensation Pools/Eco Accounts in Germany

Clearly, compensation measures[1] must be effective in order to achieve the ecological aims of the IMR. One main element of the IMR is to counterbalance negative impacts on nature and landscape with long-term positive ecological effects. Therefore, compensation should be organized in the most effective and sustainable way possible.

In Germany, compensation pools (also sometimes called eco-accounts, or in the international context related to the term habitat banks, see below for a definition) are currently employed as an important instrument for effective compensation. The concept of compensation pools is closely linked to the establishment of compensation agencies, who are responsible for creating and maintaining the pools.

"Pooling" compensation measures was first introduced into the methodology of the IMR around the year 2000. In ensuing years it has become widely adopted. Today, several compensation agencies have been active for more than ten years while their umbrella association, the *Bundesverband der Flächenagenturen in Deutschland e.V.* (*BFAD*) celebrated its tenth birthday in 2016.

The ideas of "pooling" and "banking" compensation measures were motivated by some practical shortcomings in the IMR's practice as well as discussion about the possibility of increasing flexibility. Some drawbacks of the previous system were:

- The difficulty in finding appropriate sites for effective compensation measures, resulting in compensation targeted on areas that were randomly available rather than on sites selected by appropriate ecological criteria.
- Slow approval procedures and project realization due to the lack of suitable sites for compensation measures.
- The poor implementation and, particularly, maintenance of some previously approved compensation measures, as revealed by empirical studies (see Tischew et al. 2010). Some measures stipulated in project permits were not realized at all.

[1] In this chapter, the term "compensation measure" encompasses both compensation (*Ausgleich*) and offsetting (*Ersatz*) measures unless otherwise stated.

Table 2 Examples of typical measures Bayerisches Staatsministerium für Landesentwicklung und Umweltfragen (2003, pp. 33–34, altered and translated)

Baseline situation	Measure
Habitat and land use types to be created over the short to medium term	
Farmland, grassland, fallow land	Single trees, tree rows, groups of trees and tree-lined avenues, development stages of hedges, bushes, groves and forest edges, young orchard meadow
Hypertrophic and eutrophic standing water bodies, farmland, grassland, opencast mining areas	Wetland biotopes (e.g. ponds, land-forming areas, reed belts, reed sweet-grass and bulrush communities)
Reinforced sources, degraded source areas	Renaturated sources and headwater fields
Piped watercourses, drainage pipes	Non-reinforced ditches, renaturated river sections
Shore areas kept free of vegetation	Reedbeds, tall forb communities, shoreline woods, unused shoreline strips
Wet areas used for agriculture	Reedbeds, flooded meadows
Farmland in valley sites	Extensively used grassland, fresh to moist sites
Farmland, grassland, opencast mining areas	Ruderal areas, succession areas
Opencast mining areas; otherwise no typical initial biotopes	Rocky and bare-ground biotopes: cairns and rock embankments, sandy, gravely and crushed rock areas
Habitat and land use types to be created over the long term	
Farmland, grassland	Development stages of deciduous and mixed forests of particular ecological value featuring characteristic species inventory
Farmland, grassland, fallow land	Species and structure-rich hedges, bushes, groves and forest edges
Farmland, grassland, orchards	Low-nutrient or extensive grassland, orchard meadows
Streams degraded by shoreline or bed reinforcement, piping or regulation	Natural stream and river segments
Small-scale (manual) peat digs	Regeneration stages of bog-typical communities
Damp grassland sites, stream banks	Development stages of tall forb communities with characteristic species inventory
Farmland and grassland on shallow soils; fallow or shrub-covered low-nutrient or semi-arid grasslands	Development stages of sandy low-nutrient or semi-arid grasslands with characteristic species inventory
Intensively farmed wet meadows	Development stages of moor grass and nard grass meadows, sedge or rush rich damp or wet meadows, featuring characteristic species
Initial biotopes within the habitat network system specific to the animal species concerned	Habitats for the settlement of Red List vertebrates or highly endangered animals in areas where introduction is sufficiently promising due their position in the habitat network
Examples of high-quality habitat types in which compensation measures are usually excluded;	

(continued)

Table 2 (continued)

Baseline situation	Measure
impacts are normally prohibited due to their magnitude and/or difficulty with restoration/ creation	
Habitats of high ecological value: deciduous and mixed forests, bog, swamp, marsh and floodplain forests, forests and scrublands of dry warm locations, ravine forests, block and colluvium forests of high ecological value	
Transitional peat bogs and raised bogs	
Fens and wet meadows	
Natural and near-natural river and stream sections and land-forming areas of standing waters	
Low nutrient meadows, heathland, nard-grassland, open inland dunes, thermophilic margins	
Extra-alpine rock fields	

The key to finding new solutions to these problems was a more flexible approach to the spatial aspect of compensation. From the outset, the IMR allowed for the realization of offset measures (*Ersatzmaßnahmen*) at a different site to the impact location [yet normally inside the same "nature area" defined by eco-geographical maps (e.g. Scholz 1962)]. However, changes to the building law around 1998 introduced new options, such as:

- Establishing project areas ("pools") especially selected for compensation (the spatial aspect) and
- Designing compensation that makes use of habitat banking for various impacting projects (the aspect of accounting or banking)

The difference between "compensation pools" and "eco accounts" can be defined with regard to these two aspects, although today the terms are used almost interchangeably. The use of one over the other is more a question of region and "tradition" than of functional difference.

The pooling approach led to a twofold and, in the beginning, sometimes controversial discussion about possibilities of flexibilization.

From the perspective of project developers and the administrations or local authorities responsible for project permission, the most important issue was to mainstream decision-making and compensation design processes. They advocated methodological and practical improvements to make compensation measures more easily realizable and possibly more cost-effective.

Although many conservationists agreed that practical improvements in compensation were necessary and desirable, they saw some risks in a more flexible handling of the IMR. For example:

- They argued that an easier availability of compensation measures could lead to a weakening of the mitigation hierarchy, especially the avoidance principle.
- Some experts feared compensation could become an instrument to help realize ecological projects which otherwise lacked funding. It could therefore become desirable instead of being the last step of the mitigation hierarchy—at least at the sites, where pools were planned. This could cause even more problems in ensuring the avoidance of unnecessary impacts.

These critical voices can be encapsulated by Breuer's dictum "For nature conservation something, somewhere, sometime" (Breuer 2000, title translated into English), and similarly applies to the current international debate on biodiversity offsets (e.g. Pilgrim and Bennun 2014).

It was quickly recognized that dedicated institutions were required to put ideas of compensation pools and eco accounts into practice by developing and maintaining these kinds of compensation projects. The most problematic aspect of the "usual" handling of the IMR was arguably the lack of long-term stewardship for compensation. The establishment of compensation agencies under different names and in different organizational forms remedied these deficiencies: Clarified responsibilities and enhanced ecological effectiveness where reported in a lot of case examples at conferences of the BFAD and other organisations. Unfortunately, scientifical evidence has yet hardly been published outside of conference websites and grey literature.

To safeguard the procedural and ecological qualities of compensation pools in Germany, a number of regulations and guidelines have been developed in the federal states. Regardless of their specific differences, they normally include the following requirements:

- Procedural and frequently ecological criteria for the approval of compensation measures in pools and eco accounts;
- Integration of pools into approval procedures;
- Defined responsibilities and obligations of compensation agencies (e.g. reporting to authorities).

Some regulations include the option to transfer the responsibility held by project permit holders for the compensation of their impacts entirely and permanently to compensation agencies.

Table 3 only refers to regulations. In addition, there are numerous guidelines and recommendations underpinning these regulations or giving practical orientation, both in states with regulations as well as those that have not yet introduced them.

Parallel to the development of regulations by authorities, the association of compensation agencies (BFAD) has adopted a voluntary quality standard, which has been adopted by most agencies working at a regional level. It consists of five points:

1. Compensation pools have to deliver ecological improvements. It is insufficient merely to maintain an existing level, even of high ecological value.

Table 3 List of regulations for compensation pools/eco accounts in Germany's constituent states

Federal state	Regulation/Decree
Baden-Württemberg	Eco account regulation (*Ökokonto-Verordnung – ÖKVO*) of 19 December 2010
Bavaria	Compensation regulation (*Bayerische Kompensationsverordnung – BayKompV*) of 7 August 2013
Brandenburg	Compensation pool regulation (*Flächenpoolverordnung -FPV*) of 24 February 2009
Berlin	None
Bremen	None
Hamburg	Eco account regulation (*Ökokontoverordnung – ÖkokontoVO*) of 3 July 2012
Hessen	Compensation regulation (*Kompensationsverordnung – KV*) of 1 September 2005
Lower Saxony	None
Mecklenburg-West Pomerania	Eco account regulation (*Ökokontoverordnung – ÖkoKtoVO M-V*) of 22 May 2014
North Rhine-Westphalia	Eco-account regulation (*Ökokonto VO*) of 18 April 2008
Rhineland Palatinate	None
Saarland	None
Saxony	Eco account regulation (*Ökokonto-Verordnung –SächsÖKoVO*) of 1 August 2008
Saxony-Anhalt	Eco-account regulation (*Ökokonto-Verordnung*) of 21 January 2005
Schleswig-Holstein	Eco account and compensation register regulation (*Ökokonto- und Kompensationsverzeichnisverordnung– ÖkokontoVO*) of 23 May 2008
Thuringia	None

2. Sites and measures in compensation pools must be secured for long periods of time: sites normally for an unlimited period and measures for at least 25–30 years.[2] This must be ensured by appropriate treaties and requires financial backing.
3. Proper documentation of the conditions of pools before the project started up to the effects of the measures (for example effects after 5, 10, 15 and 25 years).
4. Regional landscape planning and other relevant ecological plans and programs have to be taken into account (see von Haaren et al. 2008).
5. The high quality of planning procedures and documents has to be guaranteed, normally by hiring qualified consultants and experts from the region.

[2]25–30 years is about active management measures, their funding and safeguarding in responsibility of those who cause the impacts. For this, 25–30 years is not a short option. After this period, there might be an ecologically valuable site with conditions which of course will need further care or at least longer protection.

2 Practice in Germany

2.1 *Case Study: The Guidelines of the Federal State of Saxony*

The following figures and tables illustrate the method to calculate impact and the ecological credit points in order to restore and offset the impact of a hypothetical road building project in Saxony. This follows the Federal State of Saxony's Implementation Guidelines for Impact Mitigation Regulation (Fig. 1).

The hypothetical project is to construct a new street which will affect the existing biotopes through soil sealing and the introduction of verges at the road shoulder. For instance, oak-hornbeam woodland with a biotope value of 27 points per ha will be completely destroyed by roadwork over an area of 1.35 ha (see also sub-area 1 of Table 4). Subsequently, the roadway would only count for 0 points of biodiversity value, resulting in a net loss of −27 points per ha and −36.45 points over the impacted area. The woodland is also destroyed by the greened road shoulder over an area of 0.8 ha. However, as the biotope type 'grass verge' is assigned a residual value of 5 points per ha according to Saxony's biotope value table, the net loss is −22 points per ha, leading to a total loss over the impacted area of −17.6 points. These impacts on the oak-hornbeam woodland result in a total requirement for offset measures of at least 54.05 points in order to achieve no net loss of biodiversity value.

The considerable age of the woodland biotope type prevents an in-kind and on-site implementation, so that real biodiversity offset measures (off site and out-of-kind) become necessary. In the case of other interventions, for instance affecting the second biotope type 'other damp grassland', restoration compensation measures can be implemented in-kind and on-site. This hypothetical example also illustrates the distinction between 'restoration compensation' and real 'biodiversity offset' measures. Table 4 shows the sum of all required offset scores according to the different sub-areas and according to the different biotope types. The final analysis reveals the need for 45.25 scoring points for on-site and in-kind restoration compensation measures and 54.94 points for real biodiversity offset measures (out-of-kind and off site). This gives a concrete impression of the standard approach to biotope valuation.

The impact is compared with the compensation and biodiversity offset measures. This is shown in the following Fig. 2 with the corresponding calculation Tables 5 and 6.

The measure 'biodiversity offset 2' comprises the re-establishment of the small canal-like stream to a near-natural watercourse in an area of 0.5 ha (see Table 6 biodiversity offset no. 2). A watercourse that is not near-natural has a biotope type value of 10 points, whereas a near-natural watercourse has 27 points. This results in a potential value increase and improvement of 17 points. Multiplying by the area of 0.5 ha yields a total value increase of 8.5 value units for this measure. The measure 'restoration compensation 1' comprises a change from intensive arable land use to

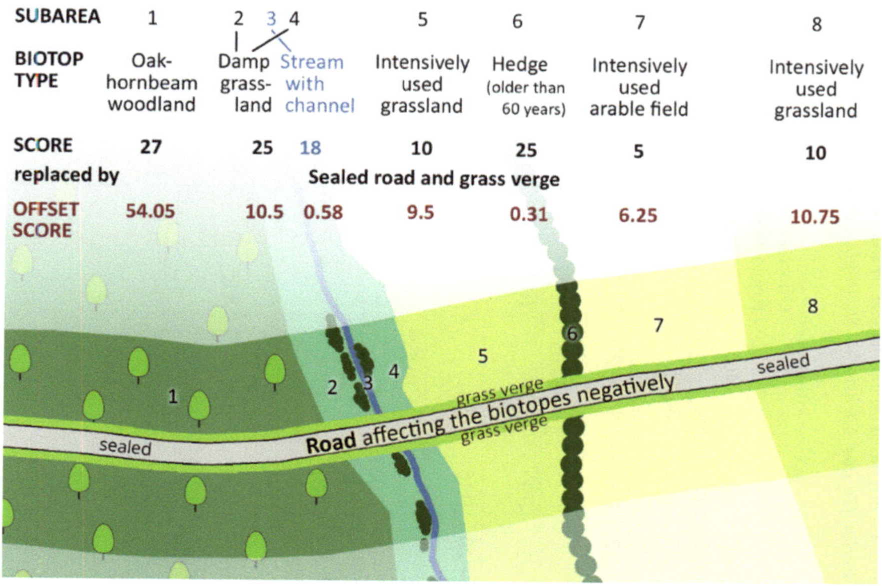

Fig. 1 Case study of the implementation guidelines for impact mitigation regulation, assessment map/*Fallbeispiel zur Handlungsempfehlung Eingriffsregelung* (altered and translated by IOER based on the source: TU Berlin for the Saxony State Ministry of the Environment and Agriculture/*geändert und übersetzt durch Leibniz-IÖR nach der Quelle TU Berlin für das Sächsische Staatsministerium für Umwelt und Landwirtschaft* 2003)

grassland with the establishment of wetland patches (see Table 5 restoration compensation no. 1). Intensively used farmland has a biotope type value of 5 points per ha, while species-rich grassland with wetland patches has a biotope type value of 22 points per ha. This yields a potential value increase of 17 points per ha. Multiplying by the area of 1.1 ha yields a total value increase of 18.7 value units. The measure ‘restoration compensation 2’ aims to change arable land use into de-intensified mesophile grassland. Whereas intensively used farmland has a value of 5 points per ha, de-intensified mesophile grassland has a value of 22 points per ha, resulting in a further potential value increase of 17 points per ha. Multiplying by the area of 1.25 ha yields a total value increase of 21.25 points through this measure. The measure ‘biodiversity offset 1’ encompasses afforestation of oak-hornbeam woodland on farmlands. Farmland has 5 credit points per ha and oak-hornbeam afforestation 23 points per ha, providing a potential value increase of 18 points per ha. Multiplying by the planned afforestation area of 2.7 ha yields a total value increase of 48.6 credit points. The measure ‘restoration compensation 3’ includes the de-intensification of existing grassland (e.g. by reducing the use of fertilizer and only mowing once a year). Intensive grassland has a biotope type value of 10 points per ha, whereas de-intensified mesophile grass land has a value of 22 points per

Table 4 Pre-intervention value of biotopes and value of biotope loss (altered and translated by the IOER based on the analysis by the TU Berlin for the Saxony State Ministry of the Environment and Agriculture/*Leibniz-IÖR nach der Quelle TU Berlin für das Sächsische Staatsministerium für Umwelt und Landwirtschaft* 2003)

	Negatively affected biotopes			Development							
Sub-area	Biotope code	Biotope type	Score	Biotope code	Biotope type	Score	Net score	Area (ha)	Net value loss	Offset score requirement (in-kind and on site)	Offset score requirement (out-of-kind, off site)
1	7 – WLE	Oak-hornbeam woodland	27	9 5 100	Road (sealed)	0	−27	1.35	−36.45	–	54.05
				9 5 600	Grass verge	5	−22	0.8	−17.6		
									∑ − 54.05		
2	4 – GFY	Other damp, species-rich grassland (presence of rare species such as *Orchis mascula*)	25	9 5 100	Road (sealed)	0	−25	0.3	−7.5	10.5	–
				9 5 600	Grass verge	5	−20	0.15	−3		
									∑ − 10.5		
3	*03220*	Stream with straightened channel/artificial banks and semi-natural elements	18	9 5 100	Road (sealed)	0	−18	0.025	−0.45	–	0.58
				9 5 600	Grass verge	5	−13	0.01	−0.13		
									∑ − 0.58		
4	4 – GFY	Other damp, species-rich grassland (presence of rare species such as *Orchis mascula*)	25	9 5 100	Road (sealed)	0	−25	0.25	−6.25	8.25	–
				9 5600	Grass verge	5	−20	0.1	−2		
									∑ − 8.25		

5	06320	Intensively used permanent mesic grassland	10	9 5 100	Road (sealed)	0	−10	0.8	−8	9.5	–
				9 5 600	Grass verge	5	−5	0.3	−1.5		
									∑ − 9.5		
6	6 5100	Hedge more than 60 years old	25	9 5 100	Road (sealed)	0	−25	0.01	−0.25	–	0.31
				9 5 600	Grass verge	5	−20	0.003	−0.06		
									∑ − 0.31		
7	10120	Intensively used arable field	5	9 5 100	Road (sealed)	0	−5	1.25	−6.25	6.25	–
				9 5 600	Grass verge	5	0	0.4	0		
									∑ − 6.25		
8	06320	Intensively used permanent mesic grassland	10	9 5 100	Road (sealed)	0	−10	0.8	−8	10.75	–
				9 5 600	Grass verge	5	−5	0.55	−2.75		
									∑ − 10.75		
										∑ 45.25	∑ 54.94

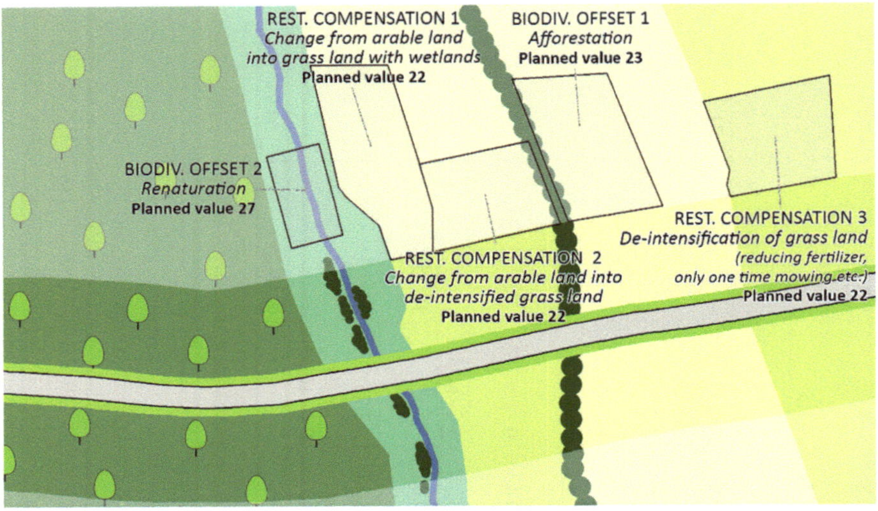

Fig. 2 Hypothetical compensation and offset—case study for the implementation guidelines for impact mitigation regulation (altered and translated by IOER, based on an analysis by the TU Berlin for the Saxony State Ministry of the Environment and Agriculture)

ha. This yields a potential value increase of 12 points per ha. Multiplying by the measure area of 0.5 ha yields a total value increase of 6 credit points.

The restoration compensation measures thus yield a total increase in value of 45.95 points. The biodiversity offset measures offer a total benefit of 57.1 points. This is to be compared with the impacts and the resulting requirement of 45.25 points for on-site and in-kind restoration compensation and 54.94 points for real biodiversity offset measures. This results in an approximate no net loss for the biotope-related approach.

However, this approach is not sufficiently comprehensive in that, so far, it has only dealt with biotopes while ignoring impacts on the ecosystem functions, which must also be compensated. Thus, the extent of the interventions into the retention function of the floodplain, into the biotic yield function and into the aesthetic function (i.e. the aesthetic qualities of the landscape) must also be calculated, compensated or replaced by suitable measures. For example, the restoration of stream floodplains to a natural state serves to increase the retention function; afforestation serves to compensate for the biotic yield function; and the dismantling of a technical structure or a steel mast serves as a form of aesthetic compensation. Thus, the approach is not narrowly focused on biotopes and/or biotope types but also addresses ecosystem functions and services.

Table 5 Biotope type value loss and related restoration compensation—case study for the implementation guidelines for impact mitigation regulation (altered and translated by IOER, based on the analysis by the TU Berlin for the Saxony State Ministry of the Environment and Agriculture)

Meas. No. (1 to x)	Biotope code	Measure (A = initial biotope; Z = target biotope)	Initial value	Planned value	Difference value	Area [ha]	Offset score (in-kind and on site)
Rest. comp. 1	10120	A: intensively used arable field	5		17	1.10	18.7
	GFY	Z: other damp grassland (species rich)		22			
Rest. comp. 2	10120	A: intensively used arable field	5		17	1.25	21.25
	GM+GB	Z: de-intensified mesic grasslands		22			
Rest. comp. 3	06320	A: intensively used permanent mesic grasslands	10		12	0.5	6.0
	GM+GB	Z: de-intensified mesic grasslands		22			
							$\sum = 45.95$

2.2 *Case Study: North-Rhine Westphalia*

Here we discuss a further example of avoidance, minimisation as well as compensation and offset measures in the construction of a bypass road in North-Rhine Westphalia to the south-east of Münster (L 585n Wolbeck bypass). The following illustrations show the position of the preferred route, which was the subject of a planning and permission study. One focus of this study (accompanied by a landscape maintenance plan) was to avoid or minimize the environmental impact by careful planning of the route and optimisation of construction work as a result of an environmental impact study. Technical measures to reduce the environmental impact were, in particular, the provision of wildlife crossings such as larger migration tunnels as well as tunnels for amphibians. In this way the project obeyed the mitigation hierarchy, according to which environmental impacts must first be avoided or minimised before consideration of biodiversity compensation measures and offsets (Figs. 3, 4, 5, 6 and 7).

Table 6 Biotope type value loss and related biodiversity (Biotope type) offset—case study for the implementation guidelines for impact mitigation regulation (changed and translated by IOER based on the analysis by the TU Berlin for the Saxony State Ministry of the Environment and Agriculture)

Meas. No. (1 to x)	Biotope code	Measure (A = initial biotope; Z = target biotope)	Initial value	Planned value	Difference value	Area [ha]	Offset score (out-of-kind, off-site)
Biod. Offs.1	10120	A: arable field/ land	5		18	2.7	48.6
	WLE	Z: oak-hornbeam woodland		23			
Biod. Offs.2	2 1 200 2/kb	A: stream with straightened channel/artificial banks	10		17	0.5	8.5
	2 1 200 1/n	Z: near natural reinstalled stream		27			
							Σ = 57.1

The two photos (Fig. 6) illustrate an area of arable land E 10.4 (marked in light green in the map above, Fig. 5) with an artificial ditch before (left) and after (right) implementation of biodiversity offsets to ensure the near-natural restoration of the ditch as well as the transformation of the arable land into an area of biotic succession with scattered plant growth, and planting of a hedge row.

2.3 *Case Study on Habitat Banking/Compensation Pools/Eco Accounts in Brandenburg*

The federal state of Brandenburg was a forerunner in the development of compensation pools, making important contributions to the concept. Between 2001 and 2005 in the Havel river region, a model project largely financed by the German Federal Agency for Nature Conservation (BfN) tested methodological and planning tools for the implementation of pools as well as a compensation agency designed as a new type of regional actor (Jessel et al. 2006) (Fig. 8).

Since its successful establishment, the Compensation Agency Brandenburg (*Flächenagentur Brandenburg GmbH*) has realized over 20 compensation pools in nearly all natural areas (*Naturräume*) of Brandenburg. Where possible, these pools are certified by the Brandenburg Ministry of the Environment according to the procedures and criteria laid out in the Compensation Pool Regulation (*Flächenpoolverordnung*) as well as the Guidelines on Realizing the Impact

Length of the investigated area	6,444 m
Length of the bypass	5,880 m
No. of bridges	7
Area newly sealed/paved	9.53 ha
Area for compensation and offset measures	67.591 ha

Fig. 3 Aerial photo of the project site and framework data on the by-pass road project L 585 Wolbeck. (Source: Reppenhorst et al. Landesbetrieb Straßenbau NRW 2007)

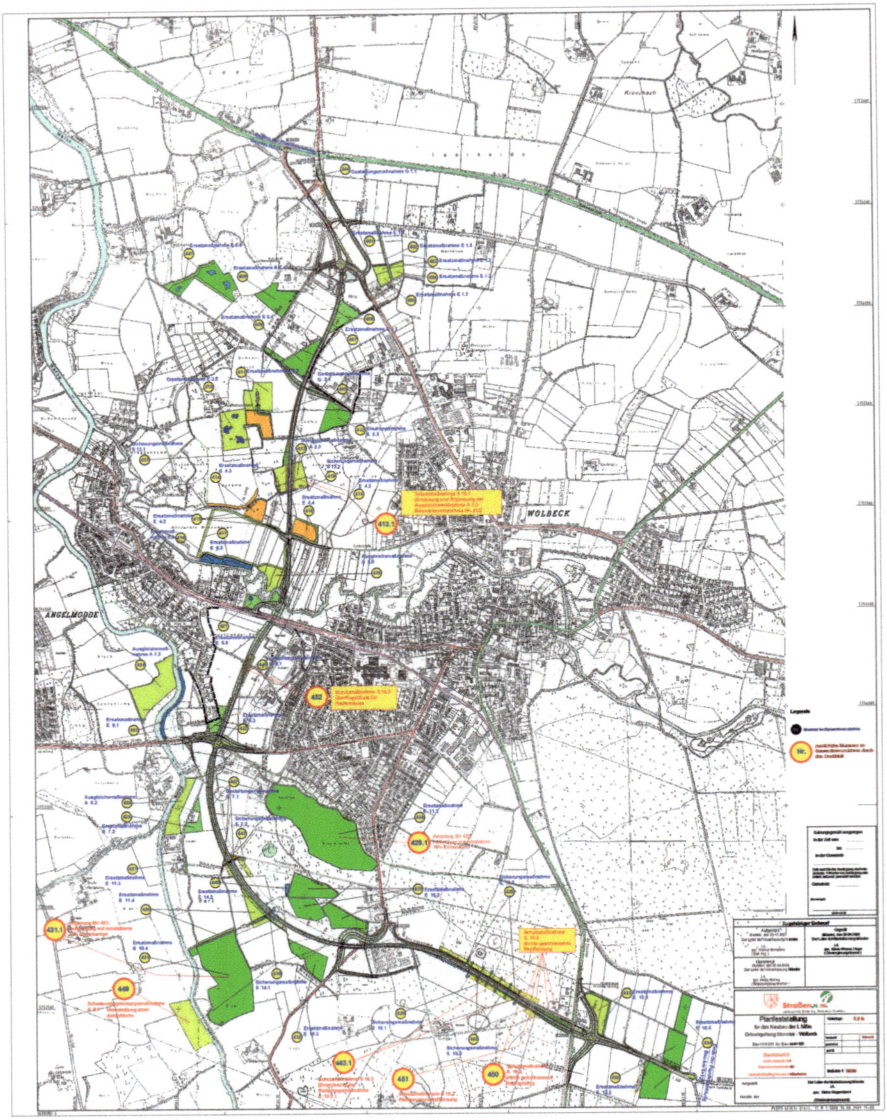

Fig. 4 Planned areas for compensation measures as well as measures to reduce and minimise environmental impacts. (Source: Reppenhorst D, Siebeneck K, Bartmann B, Landesbetrieb Straßenbau NRW 2007)

Mitigation Regulation in Brandenburg (*Hinweise zum Vollzug der Eingriffsregelung*) (Table 7).

The compensation pool "Grenzelwiesen" near the town of Beelitz (south of Berlin) is an example of a certified pool. In this (strongly altered) floodplain of the

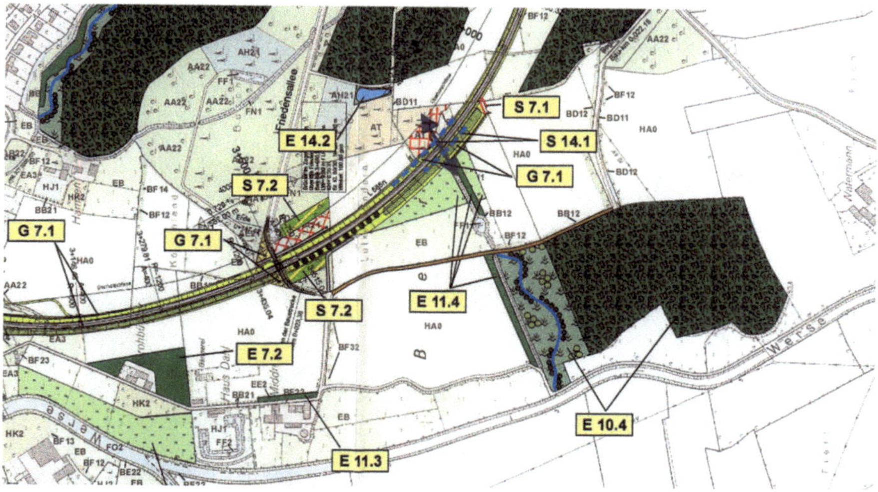

Fig. 5 Detail of the south-west planning area (rotated) from the accompanying landscape maintenance plan (Landesbetrieb Straßenbau NRW from 29 Sept. 2003)

Fig. 6 Offset measure no. E 10.4 for the bypass road Wolbeck. (Photos: Wende)

river Nieplitz, an area of about 50 ha of previously drained and intensively used agricultural land was ecologically restored to create a complex of rewetted peatlands and extensively used meadows. This project exemplifies the possibilities and the tasks connected to compensation pools:

Fig. 7 Photos showing a section of the landscape before (above) and after (below) construction of the bypass road, i.e. before and after the impact. (Photos: Wende)

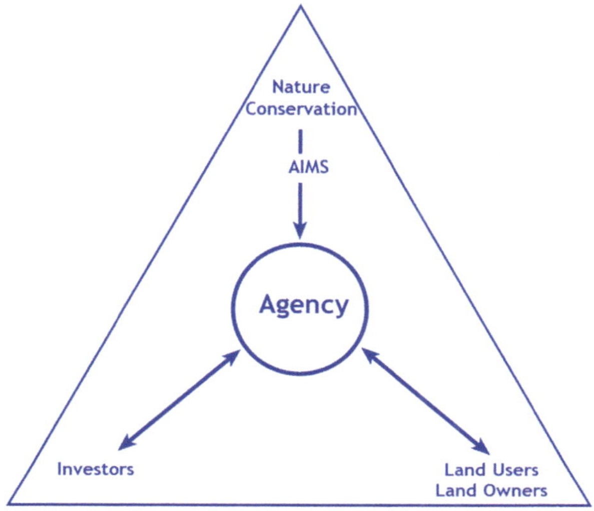

Fig. 8 Compensation agency as a hub between the different parties of the IMR (based on an idea from Arbeitsgemeinschaft Kulturlandschaft Mittlere Havel 2001)

Table 7 List of criteria for the certification of compensation pools in Brandenburg (MUGV 2009)

No.	Criterion
1	Location [which "natural area" (*Naturraum*)]
2	Detailed maps of the compensation pool site and the planned measures
3	Pool concept (with regard to landscape planning recommendations and explanations of the possible improvements in ecological functions)
4	Demand for compensation in the respective natural area
5	Coherence/extent: Pool should incorporate at least ca. 30 ha
6	Are the properties needed for the pool available?
	Do land users agree?
	→ Must be proven for at least 10 ha
7	Involvement of nature conservation administrations
8	Involvement of municipalities and their development and reference to the local landscape plans
9	Documentation on the status quo
10	Targets and control criteria for the pool
11	Commitment of the pool agency for long-term-stewardship
12	Commitment of the pool agency for periodical reports to the Brandenburg Ministry for Nature Conservation

- The ecological project pursued here was complex, large-scale and demands detailed preparation (including an approval process).
- As the land ownership structure before the project was fragmented, an important preliminary step was to acquire the land.
- The initiative came from within the region, in this case specifically from the farmer using the land.

- A long-term partnership with the farmer is essential to maintain the pool and develop its ecological potentials.
- Compensation measures from the pool where sold to over 100 clients for impacts of very different magnitudes.

Several lessons were learnt about the altered dynamics of the site during the planning phase of the pool. For example, the degradation of organic soils as a consequence of decades of drainage had changed the ground level, preventing the restoration of the previous morphology. A detailed survey of the level of the terrain had to be carried out, leading to the abandonment of some initial planning ideas.

After realisation of the measures, which largely consisted of closing draining ditches and the reintroduction of water through a new watercourse (see Fig. 9), the maintenance by de-intensified agricultural use had to be adopted to the new conditions.

The entire process of restoration and long-term site maintenance from 2011 would not have been possible without the existence of the compensation agency as a new type of project developer. Clearly, a responsible institution was required to manage the many diverse steps involved in the restoration project from land acquisition to changes in the mowing and grazing regimes.

The cooperation of the farmer was equally important. He continued to work the site after realisation of ecological measures, which were closely coordinated with him by the agency. The farmer's loss through reduced yield is made up by a 25-year maintenance treaty guaranteeing him a salary for extensive mowing and grazing. This salary makes up part of the cost of compensation measures along with the money needed for all other services necessary to realise the process (land acquisition, surveys, planning, permissions, construction work etc.), a fee for the services of the compensation agency and an appropriate risk/profit percentage.

The main message to be derived from this case study is that the success of biodiversity offsets very much depends on a strong institutional background, on long-term stewardship of target sites, on the quality of compensation as well as on monitoring by compensation agencies, and, if necessary, enforcement by authorities (Fig. 10).

3 Conclusions

Considering the development pressures on biological diversity, it is vital that we continuously deal with impacts on nature and landscapes. One third of all flora and fauna species in Germany is currently under threat, whilst approx. 6% are already extinct (BfN 2009). Furthermore, about one third of all biotopes have been classified as highly endangered (BfN 2006). Two thirds of our cultural landscapes have been fundamentally transformed by the expansion of renewable energy infrastructure as well as the growth in settlement and transport areas. It can be assumed that powerful transformative forces will continue to affect 46% of our cultural landscapes up to the

Fig. 9 Bringing the water back to the fen by building a new watercourse (2011)

Fig. 10 Before (left) and after (right) compensation measures for rewetting peatlands were realised in the compensation pool "Grenzelwiesen"

year 2030 (Schmidt et al. 2014). The average daily consumption of land is still 69 ha per day (Statistisches Bundesamt—Destatis 2017). Clearly, we are still some way from achieving the target values set by the National Strategy on Biological Diversity.

These development trends highlight the work that needs to be done to prevent the loss of biodiversity by implementing practical measures not only in protected areas but across the country. In Germany this is primarily achieved through the *Eingriffsregelung* (Impact Mitigation Regulation), which incorporates the coverage approach and the necessary regulatory tools. The Impact Mitigation Regulation—Germany's primary instrument of modern nature conservation—open up a range of options for action. These are strongly rooted in the basic principles of environmental and nature protection, namely the avoidance of unnecessary environmental impacts, the polluter-pays principle and the obligation to provide physical compensation. Reflecting the various areas of application and material repercussions, Impact Mitigation Regulation provides for a comprehensive programme of consistent impact avoidance and physical compensation, also in the form of biodiversity offsets. Of course, this does not necessarily solve all problems of land consumption or habitat and landscape fragmentation. With its focus on protection, the Impact Mitigation Regulation can be well described as conservative (in both senses), or indeed reactive. Even the legally imposed requirement of ecological compensation only achieves an approximation of the original state of the landscape. Yet considering the described developmental pressures, it is absolutely essential to ensure the functional, spatial and speedy compensation of impacts through practical measures to upgrade and restore landscapes.

Society's perception of the instrument of Impact Mitigation Regulation has shifted over the past few years. Today, all those with a stake in nature, whether developers, other land users or nature conservationists, work together to seek good and sustainable solutions of practical compensation and biodiversity offsets. The public's attention is now focussed on a fair balancing of interests. Those who gain from the exploitation of nature and landscapes must provide compensation and offsets for negative impacts. From many sides we hear the demand that mitigation and compensation measures should also benefit those affected by a project in its spatial extent. In Germany this view reflects a high awareness of the value and benefits of the natural environment. In a nationwide survey, over 90% of respondents claimed that nature and landscapes contributed to a good life (94%), to health and relaxation (92%) as well as to the education of children (92%) and to the provision of diverse experiences (92%) (BfN 2015).

Against this backdrop, Impact Mitigation Regulation should not be merely seen as ensuring piecemeal solutions, whereby each individual blade of grass and each tree has to be compensated. Rather it comprehensively regulates the balance of nature, the shaping of landscapes and biodiversity. In this way, the Impact Mitigation Regulation affects a much wider range of social activities.

Impact Mitigation Regulation is able to relieve pressures in other sectors, for example in the shaping of landscapes, in promoting leisure activities, in balancing the agricultural sector and much more. In so doing it fosters a cooperative approach between actors. These considerations once again emphasize the fact that we need a permanent nationwide principle of physical compensation and offsetting, one which must be more consistently applied in order to meet current challenges such as the expansion of the electricity grid. Here Germany's long established and sophisticated system of habitat banking of land and measures offers a wealth of sustainable concepts.

Germany's Impact Mitigation Regulation encompasses a wide range of methodological approaches and practical instruments. Yet such diversity of approach brings risks as well as opportunities. The range of application is good if this means that better account is taken of unique ecological and regional features. On the other hand, the diversity of approach becomes a problem if this leads to disparate treatment of habitats, species or landscapes. And it can seed confusion amongst investors, planners, conservationists and regulatory authorities.

Hence, if the Impact Mitigation Regulation were further clarified and standardised nationwide, this would improve the conditions for investment, speed up administrative proceedings, increase the transparency of decisions by regulatory agencies and raise the planning and legal security of private and public projects. It would also serve the interests of nature protection and landscape conservation, as an improved basis for decision-making can only increase the acceptance of measures and facilitate their long-term implementation.

We are certain to need national nature protection solutions in the future. For example, the national development plan *Netz* (2013) forecasts an extra 7900 km of new and extended lines for the national power grid, whereas the *Federal Transport Infrastructure Plan 2030* predicts an extra 8300 km of new and expanded roadways. Challenges to society such as climate change and adaptation to its repercussions, the expansion of renewable energies, the protection of wetlands and environmentally-friendly flood protection, the implementation of the EU's Water Framework Directive, the establishment of green infrastructure (cf. BfN 2017) as well as integrated urban development all require answers that must accommodate a fair balance of interests. At the same time, land users and the wider public have a legitimate demand for more cooperation and participation in decisions.

The goals of nature protection and landscape conservation can certainly be realised through the principles of the Impact Mitigation Regulation, which are the avoidance of impacts and the provision of physical compensation and offsets. In view of the major transformations that are forecast for the coming years, we require a strong and viable instrument to generate sufficient land for the methodological and practical requirements of nature protection and landscape conservation. One indispensable element is physical compensation, as this deals directly with the damaged functions of nature, landscape and biological diversity. A system that only applies compensatory payments will not lead to sustainable results.

Summarizing, we can say that Germany's Impact Mitigation Regulation is not merely a local solution to a local problem. Rather, by obeying the principle of the *avoidance* of intervention, it is anchored in the United Nations' Convention on Biodiversity. Furthermore, the Impact Mitigation Regulation closely reflects the "mitigation hierarchy" adopted by European and international regulatory systems (see also principles in chapters "Introduction of a European Strategy on No Net Loss of Biodiversity" and "Principles of the Business and Biodiversity Offsets Programme"). The long-term physical compensation and offsetting of impacts delivered by the IMR may therefore be a successful model for the conservation of biodiversity elsewhere in Europe, especially in countries dominated by highly modified habitats and with similar development pressures, and physical compensation and offset for impacts is a successful model for the preservation of biodiversity.

References

Albrecht J, Schumacher J, Wende W (2014) The German impact-mitigation regulation – a model for the EU's no-net-loss strategy and biodiversity offsets? Environ Policy Law 44(3):317–325

Anhuf D, Bräuning A, Frenzel B, Stumböck M (2003) Die Vegetationsentwicklung seit dem Ende der letzten Eiszeit. In: Kappas M, Menz G, Richter M, Treter U (eds) Nationalatlas Bundesrepublik Deutschland. Klima, Pflanzen- und Tierwelt, vol 3. Spektrum Akademischer Verlag, Berlin, pp 88–91

Arbeitsgemeinschaft Kulturlandschaft Mittlere Havel (IUS GmbH & PAN GmbH) (2001) Entwicklung und modellhafte Umsetzung einer regionalen Konzeption zur Bewältigung von Eingriffsfolgen durch einen Ausgleichspool am Beispiel der Kulturlandschaft Mittlere Havel. BfN-Skripten 37, Bonn, p 92

Balla S, Brauns F, Herberg A, Pufahl A, Schkade M et al (2000) Die Umsetzung von Kompensationsmaßnahmen der Eingriffsregelung in der Bauleitplanung. Natur und Landschaft 75(4):137–144

BauGB (1960) Baugesetzbuch vom 23.06.1960, in der Fassung der Bekanntmachung vom 3. November 2017, Federal Law Gazette I, p 3634

BauROG (1997) Bau- und Raumordnungsgesetz vom 18. 8. 1997, Federal Law Gazette I, p 2081

Bayerisches Staatsministerium für Landesentwicklung und Umweltfragen (2003) Bauen im Einklang mit Natur und Landschaft. Eingriffsregelung in der Bauleitplanung. Ein Leitfaden, vol 2. München

BKompV (2013) Bundesministerium für Umwelt, Naturschutz, Bau und Reaktorsicherheit 2013 Verordnung über die Kompensation von Eingriffen in Natur und Landschaft, Bonn, 19 April 2013. http://www.bmub.bund.de/fileadmin/Daten_BMU/Download_PDF/Strategien_Bilanzen_Gesetze/Kompensationsverordnung/entwurf_bkompV_19-04-13_bf.pdf. Accessed 5 Oct 2016

BNatSchG (1976) Gesetz über Naturschutz und Landschaftspflege (Bundesnaturschutzgesetz – BNatSchG) vom 20. Dezember 1976. Federal Law Gazette I, p 3574

BNatSchG (2002) Gesetz über Naturschutz und Landschaftspflege (Bundesnaturschutzgesetz – BNatSchG) vom 25. März 2002, Federal Law Gazette I, p. 1193

BNatSchG (2009) Bundesnaturschutzgesetz vom 29. Juli 2009, Federal Law Gazette I, p 2542

Breuer W (2000) Die Eingriffsregelung in der Bauleitplanung- Für den Naturschutz irgendwas irgendwo irgendwann? Landschaftsplanung.NET 1/2000. www.landschaftsplanung.net/texte/01_00/breuer/breuer.pdf

Bruns E (2007) Bewertungs- und Bilanzierungsmethoden in der Eingriffsregelung. Analyse und Systematisierung von Verfahren und Vorgehensweisen des Bundes und der Länder. https://depositonce.tu-berlin.de/bitstream/11303/1906/1/Dokument_16.pdf

Bundesamt für Naturschutz (2006) Rote Liste der gefährdeten Biotoptypen Deutschlands. Zweite fortgeschriebene Fassung 2006. Naturschutz und Biologische Vielfalt 34

Bundesamt für Naturschutz (2009) Rote Liste gefährdeter Tiere, Pflanzen und Pilze Deutschlands. Band 1: Wirbeltiere. Naturschutz und Biologische Vielfalt 70(1):386

Bundesamt für Naturschutz (2017) Bundeskonzept Grüne Infrastruktur. Grundlagen des Naturschutzes zu Planungen des Bundes. Bundesamt für Naturschutz, Bonn

Bundesamt für Naturschutz und Bundesministerium für Umwelt, Naturschutz, Bau und Reaktorsicherheit (2015) Naturbewusstsein 2015. Bevölkerungsumfrage zu Natur und biologischer Vielfalt. BfN Bonn und BMUB Berlin

BVerwG (Federal Administrative Court) (1990) Ruling of 27.9.1990, 4 C 44.87 (reference number)

Darbi M, Tausch C (2010) Loss-gain calculations in German impact mitigation regulation. Forest Trends, Washington. http://www.foresttrends.org/documents/files/doc_2404.pdf

Darbi M, Ohlenburg H, Herberg A, Wende W (2010) Impact mitigation and biodiversity offsets – compensation approaches from around the world. A study on the application of Article 14 of the CBD. Naturschutz und Biologische Vielfalt 101

Dickie I, Mesquite S, Morandeau D, Stoeckel M-E, Weaver J, Wende W (2013) Glossary working group on no net loss of ecosystems and their services. European Commission, DG Environment.

http://ec.europa.eu/environment/nature/biodiversity/nnl/pdf/NNL_Glossary.pdf. Accessed 21 Jun 2016

Dolde K-P (1993) Bebauungsplan und naturschutzrechtliche Eingriffsregelung. In: Driehaus H-J, Birk H-J (eds) Baurecht – Aktuell. Festschrift für Felix Weyreuther. Carl Heymanns, München, pp 195–212

Gaentzsch G (1990) Bauleitplanung und Baugenehmigungspraxis unter den Anforderungen des Naturschutzes und der Landschaftspflege. Natur und Recht 12(1):1–8

GG (2006) Gesetz zur Änderung des Grundgesetzes vom 28.8.2006. Federal Law Gazette I, p 2034

Götze R, Müller W (2008) Das Gesetz zur Erleichterung von Planungsvorhaben für die Innenentwicklung der Städte ("BauGB 2007") – Zu Risiken und Nebenwirkungen eines Planungserleichterungsgesetzes. Zeitschrift für Umweltrecht 19(1):8–16

Herbert M (2015) German impact mitigation regulation – national, European, international. A comparison from the viewpoint of the Federal Agency for Nature Conservation UVP-report 29 (3):149–151

Herbert H, Mayer F (2007) Die Eingriffsregelung heute. In: 30 Jahre naturschutzrechtliche Eingriffsregelung – Bilanz und Ausblick. Schriftenreihe des Deutschen Rates für Landespflege 80:17–21

IWG (1993) Gesetz zur Erleichterung von Investitionen und der Ausweisung und Bereitstellung von Wohnbauland vom 22. 4. 1993. Federal Law Gazette I, p 466

Jessel B (2003) Die Neufassung der naturschutzrechtlichen Eingriffsregelung nach §§ 18, 19 BNatSchG. Eine Diskussion der wesentlichen Änderungen und möglicher Auswirkungen auf exemplarische Handlungsfelder. Naturschutz und Landschaftsplanung 35(4):119–125

Jessel B, Schöps A, Gall B, Szaramowicz M (2006) Flächenpools in der Eingriffsregelung und regionales Landschaftswassermanagement als Beiträge zu einer integrierten Landschaftsentwicklung am Beispiel der Mittleren Havel. Naturschutz und Biologische Vielfalt 33:410

Koch H-J (2010) Die naturschutzrechtliche Eingriffsregelung. In: Kerkmann J (ed) Naturschutzrecht in der Praxis, 2nd edn. Lexxion Verlagsgesellschaft, Berlin, pp 119–198

Köppel J, Peters W, Wende W (2004) Eingriffsregelung, Umweltverträglichkeitsprüfung, FFH-Verträglichkeitsprüfung. UTB, Stuttgart

Landesbetrieb Straßenbau NRW (2003) Entwurf Landschaftspflegerischer Begleitplan Ortsumgehung L585 Wolbeck. Bearbeitendes Büro Froehlich & Sporbeck

Lau M (2011a) Die naturschutzrechtliche Eingriffsregelung (Teil 1). Natur und Recht 33 (10):680–684

Lau M (2011b) Die naturschutzrechtliche Eingriffsregelung (Teil 2). Natur und Recht 33 (11):762–771

Makala M (2016) Landschaftspflegerische Begleitplanung. In: Riedel W, Lange H, Jedicke E, Reinke M (eds) Landschaftsplanung, vol 3. Springer Spektrum, Heidelberg

Meyhöfer T (2000) Ausgleich und Ersatz in Bebauungsplänen: Umsetzungsdefizite, Ursachen und Lösungswege. Naturschutz und Landschaftsplanung 32(11):325–328

MUGV (Ministerium für Umwelt, Gesundheit und Verbraucherschutz Brandenburg) (2009) Hinweise zum Vollzug der Eingriffsregelung – HVE. http://www.mlul.brandenburg.de/cms/media.php/lbm1.a.3310.de/hve_09.pdf

OVG Münster (Higher Administrative Court of Münster) (1998) Ruling of 10.02.1998, 10 B 2439/97 (reference number)

Pilgrim J-D, Bennun L (2014) Will biodiversity offsets save or sink protected areas? Conserv Lett 7 (5):423–424

Rayment M, Haines R, McNeil D, Conway M, Tucker G, Underwood E (2014) Study on specific design elements of biodiversity offsets: biodiversity metrics and mechanisms for securing long term conservation benefits. http://ec.europa.eu/environment/nature/biodiversity/nnl/pdf/Biodiversity%20offsets%20metrics%20and%20mechanisms.pdf. Accessed 21 Jun 2016

Reppenhorst D, Siebeneck K, Bartmann B (2007) Landesbetrieb Straßenbau NRW, Presentation 13 Jun 2007

Scheidler A (2010) Die naturschutzrechtliche Eingriffsregelung im BNatSchG 2010. Zeitschrift für Umwelt- und Planungsrecht 30(4):134–141
Schmidt C, Hofmann M, Dunkel A (2014) Den Landschaftswandel gestalten. BfN und BBSR, vol 1,2,3
Scholz E (1962) Die naturräumliche Gliederung Brandenburgs. Pädagogisches Bezirkskabinett, Potsdam, p 93
Schrödter W (2015) Baugesetzbuch, 8th edn. Nomos Verlag, Baden-Baden
Schumacher J, Fischer-Hüftle P (2011) Bundesnaturschutzgesetz, 2nd edn. Kohlhammer Verlag, Stuttgart
Siebert J, Steingrube W (2000) Inszenierte Natur. In: Becker C, Job H (eds) Bundesrepublik Deutschland. Nationalatlas. Band 10: Freizeit und Tourismus. Spektrum Akademischer Verlag, Berlin, pp 40–43
Statistische Ämter des Bundes und der Länder (2016) Gebiet und Bevölkerung. http://www.statistik-portal.de/Statistik-Portal/de_jb01_jahrtab1.asp. Accessed 5 Oct 2016
Statistisches Bundesamt (2017) Nachhaltige Entwicklung in Deutschland. Indikatorenbericht 2016. Wiesbaden
Tischew S, Baasch A, Conrad M-K, Kirmer A (2010) Evaluating restoration success of frequently implemented compensation measures: results and demand of control procedures. Restor Ecol 18 (4):467–480
TU Berlin für das Sächsische Staatsministerium für Umwelt und Landwirtschaft (2003) Handlungsempfehlungen zur Bewertung und Bilanzierung von Eingriffen im Freistaat Sachsen – Fallbeispiel zur Handlungsempfehlung Eingriffsregelung. https://www.umwelt.sachsen.de/umwelt/download/natur/fallbsp.pdf. Accessed 21 Sep 2016
Tucker G, Allen B, Conway M, Dickie I, Hart K, Rayment M, Schulp C, van Teeffelen A (2013) Policy options for an EU no net loss initiative. Report to the European Commission. Institute for European Environmental Policy. http://ec.europa.eu/environment/nature/biodiversity/nnl/pdf/Policy%20Options.pdf. Accessed 23 June 2016
Von Haaren C, Galler C, Ott S (2008) Landscape planning – the basis of sustainable landscape development. Federal Nature Conservation Agency, Bonn. https://www.bfn.de/fileadmin/MDB/documents/themen/landschaftsplanung/landscape_planning_basis.pdf. Accessed 23 Jun 2016
Wagner S (2007) Ökokonten und Flächenpools. Die rechtlichen Grundlagen, Möglichkeiten und Grenzen der Flächen- und Maßnahmenbevorratung als Ausgleichsmaßnahmen im Rahmen der Eingriffsregelung im Städtebaurecht. Duncker & Humblot, Berlin
Weiland U, Wohlleber-Feller S (2007) Einführung in die Raum- und Umweltplanung. Paderborn
Wende W, Herberg A, Herzberg A (2005) Impact mitigation regulation – mitigation banking and compensation pools: improving the effectiveness of impact mitigation regulation in project planning procedures. Impact Assess Proj Appraisal 23(2):101–111

Lithuania

Vytaute Bacianskaite

The biggest enlargement in the history of the European Union occurred on 1 May 2004 with the accession of ten Central and Eastern European countries. One of these was Lithuania, a country then still battling the repercussions of involuntary incorporation into the Soviet Union for 50 years. Accession to the EU was a key plan in Lithuania's foreign policy from the moment independence was declared on 11 March 1990.[1]

Lithuania is a relatively small member state with a population of 2.9 million and total area of 65,000 km^2, of which 33.3% is covered by forest and 13% belongs to the Natura 2000 network of protected areas. Since acceding to the EU, Lithuania has gained a unique experience in developing its own environmental system.

Transposition of the EU environmental *acquis* was one of the most technically difficult challenges. More than 300 directives, regulations, decisions and recommendations had to be integrated into the national body of legislation.

The goal of this chapter is to identify and analyse the no net loss (NNL) mechanisms that currently exist in Lithuania and which build on the EU and Lithuanian legal systems. A number of technical, ecological, economic and other factors determine the compensation of biodiversity loss at national level. However, political resolve and the architecture of the country's legal system, as will be shown later, remain the most influential criteria for the effective protection of biodiversity.

Broadly speaking, there are two types of environmental evaluations relevant to NNL situations when impacts are to be assessed: *ex ante* evaluation of plans and projects, here illustrated through the case study on *Boros Schneideri*, and *ex post* evaluations of environmental damage, which will be presented by reference to case law.

[1]Gumbis and Bacianskaitė (2014).

V. Bacianskaite (✉)
Deutsche Bundesstiftung Umwelt 2015, Vilnius, Lithuania
e-mail: vytaute.bacianskaite@outlook.com

W. Wende et al. (eds.), *Biodiversity Offsets*,
https://doi.org/10.1007/978-3-319-72581-9_9

The analysis takes its lead from Article 6(4) of the EU Habitats Directive,[2] the only article setting down a mandatory requirement to compensate for unavoidable residual impacts on habitats and species of Community interest in the EU. Further examination of the Lithuanian experience in transposing the EU legislation, e.g. EIA/SEA[3] and ELD,[4] will provide an illustrative example in the context of the theoretical analysis of the potential of existing instruments at EU level to contribute to NNL.

1 Theory in Lithuania

1.1 Legal Background in Lithuania

In Lithuania a number of laws deal with the evaluation and compensation of negative impacts upon the environment, including impacts on biological diversity. Based on the level and/or the object of assessment, the evaluation is specified in different categories of laws (e.g. laws adopted by the Parliament "Seimas"), governmental resolutions and ministerial orders.

In the context of habitats, negative impacts on Natura 2000 sites are examined as an integral part of other assessment procedures (e.g. EIA, SEA). Therefore, the analysis on Article 6(3, 4) of the Habitats Directive in the context of EIA/SEA plays a key role for the evaluation of negative impacts in terms of NNL in Lithuania.

1.2 Challenges for Biodiversity Offsets and NNL with Regard to Natura 2000 in the Lithuanian Context

The map of European Ecological Network Natura 2000 sites in Lithuania[5] illustrates the core idea of the NNL approach (see Fig. 1). Coloured hatching denotes protected sites where a special legal regime is applied: green hatching indicates the 410 Special Conservation Areas (SACs) established under the Habitats Directive, while red hatching refers to the 83 Special Protection Areas (SPAs) under the

[2]Council Directive 92/43/EEC of 21 May 1992 on the conservation of natural habitats and of wild fauna and flora ('Habitats Directive').

[3]Directive 2014/52/EU of the European Parliament and of the Council of 16 April 2014 amending Directive 2011/92/EU on the assessment of the effects of certain public and private projects on the environment ('EIA') and Directive 2001/42/EC of the European Parliament and of the Council of 27 June 2001 on the assessment of the effects of certain plans and programs on the environment ('SEA').

[4]Directive 2004/35/CE of the European Parliament and of the Council of 21 April 2004 on environmental liability with regard to the prevention and remedying of environmental damage ('ELD').

[5]As provided by the State Service for Protected Areas: www.vstt.lt.

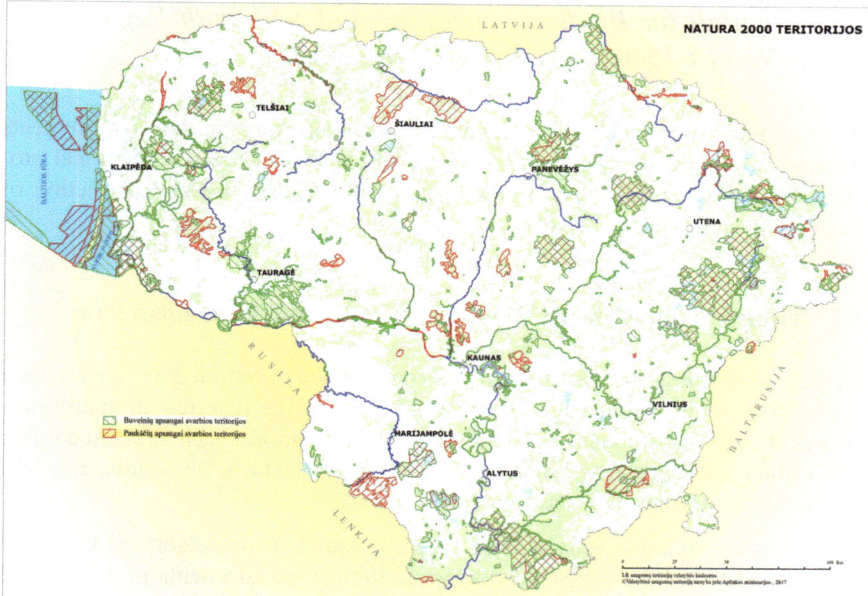

Fig. 1 The map of European Ecological Network Natura 2000 sites in Lithuania, designed by the State Service for Protected Areas: www.vstt.lt

Birds Directive in 2015. The total area of Natura 2000 sites made up 13% of Lithuania's entire territory. The key institutions responsible for the management of protected areas (including Natura 2000) are the Ministry of the Environment and the State Service for Protected Areas.[6]

Under the requirements of the Habitats Directive, the competent national authorities shall agree to the plan or project only after having ascertained that it will not adversely affect the integrity of the site concerned.[7] If a plan or project must nevertheless be carried out, all compensatory measures must be taken to ensure the overall coherence of Natura 2000. This means that a certain level of NNL in Natura 2000 territories is assured.

However, no analogous compensation requirements currently exist in regions outside Natura 2000 territory. This encourages the search for new ways of expanding NNL goals beyond specially designated areas.

[6]Ibid.

[7]Article 6(3) of the Habitat Directive.

1.3 Potentials for Biodiversity Offsets and NNL with Regard to Natura 2000 in the Lithuanian Context

Table 1 correlates the EU and Lithuanian legislation on compensatory requirements stemming from Articles 6(3) and 6(4) of the EU Habitats Directive. National provisions set the key legal basis to assure the overall coherence and protection of Natura 2000 at the national level.

1.3.1 Transposition of the EU Habitats Directive into Lithuanian Law

Although economic activities are not *a priori* prohibited in Natura 2000 sites or their neighbouring areas, Article 6(3) of the Habitat Directive requires an appropriate assessment of projects, plans or programmes that may impact such sites. According to the Lithuanian legislation,[8] there are three cases for which the significance of a planned economic activity is assessed:

1. A proposed economic activity is included in Annex 1 (mandatory EIA)[9] of the Law on EIA[10] or the developer decides to initiate the EIA without a screening procedure,
2. The planned economic activity is included in Annex 2 (screening is required)[11] of the Law on EIA, or
3. The implementation of a plan, programme or project is related to Natura 2000 sites or their neighbouring areas.

It follows that, except for programmes and plans (subject to the SEA procedure), projects not directly related to the management of Natura 2000 but which may have a significant impact on natural habitats and protected species of flora and fauna *must* be evaluated in accordance with the Law on EIA. Final impact assessment results are presented in the EIA/SEA report, as provided in the table of Articles 6(3) and 6(4) of the Habitats Directive and Lithuanian legislation (see Fig. 1).

---

[8]Order of the Minister of Environment of the Republic of Lithuania, 22 May, 2006, No. D1-255, "Approval of the Regulations of Determination of Significance of the Effects of Implementation of Plans, Programmes and Proposed Economic Activities on Established or Potential 'Natura 2000' Territories" *(Official Gazette, 2006-05-31, No. 61-2214).*

[9]List of the proposed economic activities subject to an environmental impact assessment.

[10]Law on the Environmental Impact Assessment of Proposed Economic Activity of the Republic of Lithuania No. I-1495 ('Law on EIA') [as amended by the Law No. XII-418 *(Žin., 2013-07-16, No. 76-3835)*].

[11]Ibid.

Table 1 Articles 6 (3) and 6(4) of the Habitats Directive and relevant provisions of Lithuanian laws

Council Directive 92/43/EEC of 21 May 1992 on the conservation of natural habitats and of wild fauna and flora	Lithuanian Legislation
Article 6(3) Any plan or project not directly connected with or necessary to the management of the site but likely to have a significant effect thereon, either individually or in combination with other plans or projects, shall be subject to appropriate assessment of its implications for the site in view of the site's conservation objectives. In the light of the conclusions of the assessment of the implications for the site and subject to the provisions of paragraph 4, the competent national authorities shall agree to the plan or project only after having ascertained that it will not adversely affect the integrity of the site concerned and, if appropriate, after having obtained the opinion of the general public	Law of Protected Areas of the Republic of Lithuania No. I-301 [as amended by the Law No. XII-1305 (TAR, 2014-11-13, No. 2014-16784)] Article 24^{1}. Composition and protection of Natura 2000 network 10. The projects not directly related to the management of areas of the European ecological network "Natura 2000", but which, either individually or in combination with other plans or projects, may have an adverse impact on natural habitats and protected species of flora and fauna, shall be properly evaluated <...>, taking into account the protection goals of the site. <...> the State or Municipal institutions or Agencies can approve the plan or project only after having ascertained that the plan or project will not adversely affect the integrity of the site concerned and, where necessary, upon ascertaining the public opinion Law on the Environmental Impact Assessment of the Proposed Economic Activity of the Republic of Lithuania No. I-1495 (as amended by the Law No. XII-418 (Official Gazette, 2013-07-16, No. 76-3835) Article 7. Screening and Environmental Impact Assessment 6. The competent authority shall conduct a screening procedure and make a conclusion as to whether an environmental impact assessment is required on the basis <...> having regard to: 1) the environmental sensitivity of the location likely to be affected by the proposed economic activity, the characteristics of the ecosystem, landscape, the nature of areas of used land, local infrastructure, concentration of industrial facilities, the relative abundance, quality and regenerative capacity of natural resources, the absorption capacity of the natural environment, paying particular attention to protected areas, also the environmental protection purposes of a Natura 2000 site, densely populated areas, wetlands, forest areas, protection zones, the analysis of data of performed environmental monitoring, the territories where the permissible level of pollution has been exceeded or the territories of historical, cultural or archaeological significance; <...>

(continued)

Table 1 (continued)

Article 6(4) If, in spite of a negative assessment of the implications for the site and in the absence of alternative solutions, a plan or project must nevertheless be carried out for imperative reasons of overriding public interest, including those of a social or economic nature, the Member State shall take all compensatory measures necessary to ensure that the overall coherence of Natura 2000 is protected. It shall inform the Commission of the compensatory measures adopted. Where the site concerned hosts a priority natural habitat type and/or a priority species, the only considerations which may be raised are those relating to human health or public safety, to beneficial consequences of primary importance for the environment or, further to an opinion from the Commission, to other imperative reasons of overriding public interest	Law of Protected Areas of the Republic of Lithuania No. I-301 (as amended by the Law No. XII-1305 [TAR, 2014-11-13, No. 2014-16784)] Article 24^{1} Composition and protection of the Natura 2000 network <...> 11. If, in spite of a negative assessment of the implications for the site and in the absence of alternative solutions, a plan or project referred in paragraph 10 must nevertheless be carried out for imperative reasons of overriding public interest, including those of a social or economic nature, all compensatory measures necessary to ensure that the overall coherence of Natura 2000 is protected shall be implemented. Adopted compensatory measures are communicated to the European Commission. Where the site concerned hosts a priority natural habitat type and/or a priority species, the only considerations which may be raised are those relating to human health or public safety, to beneficial consequences of primary importance for the environment. <...> Order of the Minister of the Environment of the Republic of Lithuania No. D1-959, 30 November, 2010 "Approval of the Procedure on the Application of Compensatory Measures Aimed at Preserving the Overall Integrity of the European Ecological Network Natura 2000, Notification to the European Commission about the Adopted Compensation Measures and Referral to the European Commission for the Opinion" (Official Gazette, 2010-12-07, No. 143-7345) <...> 4. Compensatory measures are special instruments of proposed economic activity, plan or programme. They are applied as an additional tool to implement the Habitats Directive (92/43/EEB) and the Directive on the conservation of wild birds (2009/147/EC). The measures are designed to offset the negative effects of the proposed economic activity, plan or programme in order to ensure the overall coherence of the ecological network Natura 2000, to compensate for the actual damage to the species and habitats 5. The compensatory measures must be oriented to the conservation objectives of the Natura 2000 site in concern, to the habitats and species that were negatively affected in similar ratio in respect of the number and status 6. Compensatory measures are applied only if it is not feasible to completely avoid adverse effects by other means and it is decided to consider the plan, programme or proposed economic activity despite its negative impact on the Natura 2000 network

(continued)

Table 1 (continued)

	7. Compensatory measures can be: 7.1. restoration of the habitat and (or) species in a new or enlarged site to be included in the Natura 2000 network; 7.2. restoration or improvement of the habitat in the same area or in another Natura 2000 site proportionate to the loss; 7.3. in exceptional cases, proposing a new site under the Habitats Directive 8. For the purposes of the Habitats Directive, the compensatory measures include the following activities: 8.1. reintroduction of species; 8.2. restoration and enhancement of species, including the intensification of nutrition (prey) species; 8.3. acquisition of land ownership or possession rights with a view to preservinge/restoring protected habitats and (or) the species; to designate a protected area; 8.4. designation of protected areas (including limitations on land use); 8.5. promotion of certain economic activities sustaining key ecological functions; 8.6. the reduction of other types of threats 9. In order to ensure the overall coherence of the Natura 2000 network, compensatory measures laid down in a plan, programme or project must: 9.1. compensate in similar proportions for the negative effects on the respective habitats and species; 9.2. be implemented in the same region (area) of the development location of a plan, programme or project in order to provide a favorable effect of the compensatory measures on the same local populations of species or in order to ensure the geographic coherence of the Natura 2000 network; 9.3. perform similar functions for which the affected area complied with the screening criteria applicable for the site of European Community Importance <…> 11. Compensatory measures must not jeopardize other sites directly related to the maintenance of the overall coherence of the Natura 2000 network 12. The costs related to the application of compensatory measures are borne by the organizer (developer) of a proposed economic activity or the organizer of a plan or a programme

1.3.2 *Ex Ante* Evaluation of Plans and Projects

Independent of the territorial status, authorization of any plan or project can be granted only if the competent authority is convinced that the environment will not be adversely affected. If a proposed economic activity falls under the scope of the Law on EIA *and* at least one component of the environment is likely to experience significant negative consequences, then the EIA must be carried out; or in the case that there is insufficient information to determine whether the consequences of the implementation of the proposed economic activity may cause significant effects.[12]

With this assumption, the EIA not only functions *de jure* as a procedural instrument. *De facto* it creates material consequences by precluding a proposed economic activity from further development if a competent authority decides that likely significant adverse effects on the environment, including biodiversity, may arise.

Where the competent authority believes that the proposed economic activity at a selected location will violate respective statutory provisions and/or is likely to have adverse effect on the environment, the proposed economic activity will not be approved.[13] If the developer is still interested in developing a particular project, s/he has to review and amend the conditions (mitigation measures, alternatives, location, etc.), thus launching a new EIA procedure. Such an approach could also serve for integrating NNL goals in an overarching *ex ante* environmental assessment mechanism for all projects and programmes in parallel to currently existing EIA/SEA regulations at the national level.

1.3.3 *Ex Post* Evaluation of Impact Under the Environmental Liability Directive

Imperative requirements for environmental compensation through the offsetting of residual impacts as determined by Article 6(4) of the Habitat Directive are extended by the Environmental Liability Directive (ELD) provisions, which require *ex post* remediation for *significant damage* inflicted on biodiversity resources and services. The ELD introduces two mechanisms to address residual impact: *complementary remediation*, to be applied when primary remediation of the affected site is not achievable, and *compensatory remediation*.[14] Implementation of these approaches varies in national systems.

The ELD provisions uphold the principle of "polluter pays" by which a developer causing environmental damage or imminent (real) threat of such damage should, in principle, be liable to reimburse the necessary costs of preventive or remedial

---

[12]Manual for Environmental Impact Assessment in Lithuania, p. 8., as provided by the Ministry of the Environment of the Republic of Lithuania: www.am.lt/en

[13]Article 10(9) of the Law on EIA.

[14]Annex II, ELD.

measures (Preamble, Paragraph 18). However, this directive should not prevent Member States from maintaining or enacting more stringent provisions in relation to the prevention and remedying of environmental damage (Preamble, Paragraph 29, Article 16).

According to the ELD, the developer must undertake the necessary remedial measures and bear the costs for any preventive and remedial actions (6, 8). Annex 2, 1-1.1.3 states that if damaged natural resources and/or services do not return to their baseline condition, complementary and compensatory measures shall be taken.

In the light of the ELD, Article 32 of the Law on Environmental Protection of the Republic of Lithuania[15] requires that the developer restores the environment to its baseline condition (if feasible) and compensates all costs. The indicated methodology for calculating the amount of compensation for environmental damage[16] precisely refers to cases where environmental contamination is irreversible, a polluter fails to restore the environment or measures are insufficient to completely restore (replace) the previous state of the environment. In such cases monetary compensation must be provided equivalent to the loss.

It is clear that both the provisions of the ELD and the Law on Environmental Protection set out the requirements for a natural or legal person causing damage to restore the damaged environment to its original state. While it is essential to restore the physical environmental restoration, additional compensations may be demanded. Thus the ELD requires that the public be compensated for the primary damage as well as the loss that occurs for the period until the environment recovers its original state.

In its ruling of 29 October, 2003,[17] the Constitutional Court of the Republic of Lithuania stated that all cases of harm (damage) inflicted on the natural environment have to be compensated, regardless whether any methods of compensation for such harm (damage) have been established. Therefore, while we can point to numerous methods of damage compensation based on categories of pollution sources, territorial status or environmental elements, the lack of a concrete methodology applicable to a particular case should not exempt a person from the constitutional duty to compensate for damage. In case of a dispute, the size of the harm (damage) may be determined according to the judicial procedure. For this reason, the competent authorities should be provided with comprehensive evidence on the facts of environmental damage and relevant information.

To conclude while similar principles of compensating damage may be found in civil law, environmental liability serves the special and singular purpose of preventing and remedying environmental damage. With its unique characteristics

---

[15]Nr. XII-2358, *TAR* 2016-13919.

[16]Order of the Minister of the Environment of the Republic of Lithuania, 9 September, 2002, No. 471 "On the Approval of the Methodology for Calculating the Amount of Compensation for Environmental Damage", (*Official Gazette 2002-09-25, Nr. 93-4026, as amended - TAR, 2014-09-01, No. 11566*).

[17]No. 1/02 "On the Procedure of the Publication of Legal Acts".

and scope, the ELD is a highly innovative legal instrument and autonomous liability regime, making a useful contribution to the NNL goals of the EU and its Member States.

1.3.4 Compensation Under the Law on Forests of the Republic of Lithuania

A special regulation on the conversion of forest can be presented as an alternative NNL-type instrument at national level. The Law on Forests of the Republic of Lithuania[18] states that forested land may be transformed into farmland or other land-use types only in exceptional cases, thereby balancing the interests of the state, the forest owner and society, and in the manner prescribed by the Government of the Republic of Lithuania. Consequently, if forested land is converted to some alternative land use, compensation has to be paid for the lost forest according to the Procedure on Forest Land Conversion and Compensation.[19]

The law is applied regardless of the form of land ownership, i.e. public or private. However, private landowners are allowed to compensate for lost woodland by replanting trees on their private land. In other cases, monetary compensation is demanded. The amount of compensation depends on several factors, such as:

(a) The market price of forested land before its conversion into other land uses. This is determined from the evaluation maps of land areas following the procedure set out in the Resolution of the Government of the Republic of Lithuania No. 1523[20];
(b) The costs of planting and maintaining woodland in the conversion area.

Compensation must be paid before deregistration of the forested land from the State Forest Cadastre. Monetary compensation goes towards the state budget revenues as well as the special forestry funding program, and is used to cover costs of a wide spectrum of forest protection and management measures implemented by the national authorities.

[18]*Official Gazette*, 2011 No. 74-3548.

[19]Resolution No. 1131 "Procedure on Forest Land Conversion and Compensation"(*Official Gazette,* 2011, *No. 120-5657).*

[20]*Official Gazette*, 2012, No. 146-7536.

2 Practice in Lithuania

2.1 Ex Ante *Evaluation of Plans and Projects: Case Study on the Potential Loss of a Beetle Habitat*

Although little information has been gathered on the application of NNL in Lithuania, several characteristic examples can help to illustrate *ex ante* environmental compensation in the light of the Habitats Directive.

In 2012 one of the largest construction companies in the Jurbarkas region, UAB Jurmelsta Ltd., submitted an application for gravel extraction in the third gravel pool of Kalnėnai. The proposed site of the economic activity was in Pašventys village, located in the west of Lithuania in the region of Jurbarkas city and within the Natura 2000 site "Karšuvos giria" (LTJUROOO8).

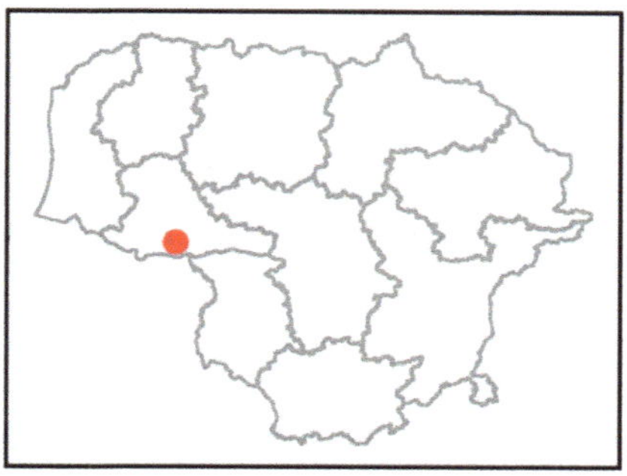

The plan was to mine gravel for road building and maintenance. The forecasted volume of extracted material was around 30,000 m^3 of unloosened gravel per annum. General mining activities were to be conducted for 9 months of the year. According to the estimated total stock of raw material, mining activities would continue for 14–15 years.

Under the general land-use zoning plan of Jurbarkas, the territory falls within a designated area for the exploitation of mineral resources. As stated by the List of Extinct and Endangered Species of Lithuania, the impact site of the proposed activity had been identified as the habitat of a protected species of beetle, Šneiderio kirmvabalis (*Boros schneideri*). If the development proceeded, it would be necessary to cut down almost 4 ha of pinewood forest, a potential habitat of *Boros schneideri*.

As a part of the EIA process, an evaluation of the project's impacts on the Natura 2000 site was carried out, particularly the impact on *Boros schneideri* (listed in Annex 2 of the Habitats Directive). A careful assessment was undertaken whether the planned project would undermine the favourable status of the beetle, which is present in the Natura 2000 site of Karšuvos giria.

Although the reasoned opinion of the State Service for Protected Areas[21] was that the proposed economic activity would not adversely affect the integrity of the site

---

[21]Based on the Decision on the Proposed Economic Activity in the 3rd Gravel Pool of Kalnėnai in Jurbarkas District, No. (PAV)-D2-1280, 29 May 2012, adopted by Kaunas Regional Department of Environmental Protection.

concerned, certain binding conditions for the project development were specified, namely:

1. To refrain from felling withered pines in the pinewood-type forest at the excavation location in Kalnėnai until their bark has fallen off;
2. To remove the brushwood in 8 ha of the forest adjacent to the impacted site.

These conditions were included in the technical project and financed by the developer at a compensation ratio of 1:2. Additionally, the developer was required to build a new recreational pond as a potential habitat for rare and protected bird species after the conclusion of the project.

Thus to summarize, despite confirming the *insignificant* negative impact on *Boros Schneideri* and the overall integrity of the Natura 2000 site (Karšuvos giria), the residual impact had to be offset in adjacent territory under the principle *like-for-like*. This required the restoration of a suitable habitat twice the size of the impacted site and within a geographically connected area.

2.2 Ex Post *Evaluation of Environmental Damage: Case Study on Monetary Compensation in Lithuania*

In the civil case No. 3K-3-61/2014,[22] the Supreme Court of Lithuania examined the relationship between the restoration of environmental damage (*restitutio in integrum*[23]) and equivalent monetary compensation. In 2009 the petroleum refinery company of the Baltic States, Orlen Lietuva, received a report regarding the illegal tapping of the "Polotsk-Ventspils" pipeline. This caused a spill of diesel fuel onto a wetland area, specifically, contaminated the Šaltoja and Vyžuonos rivers, their valleys, recognized as priority sites for bird conservation and part of the Natura 2000 network.

According to the plaintiff, the defendant did not sufficiently fulfill the duty as prescribed by the law to protect the product pipeline from unauthorized access. According to the eco-geological research report, decontamination work still left 32.3 tons of diesel fuels on the site, of which 1.89 tons were located in the Vyžuona River and its basin. Oil products flowed into the valley of the Šaltoja River, contaminating an area of 9.6 ha, of which two-thirds were wetlands. March—April, 2009, Orlen Lietuva organized neutralization works of the spill: 87 tons of diesel were collected and removed from a mixture of soil and water. Although a major part of pollutants was collected, more than one-fourth was still found in the area.

[22] *"Lietuvos Respublikos aplinkos ministerijos Panevėžio regiono aplinkos apsaugos departamentas v. AB 'Orlen Lietuva"*, No. 3K-3-61/2014.

[23] The Latin term *restitutio in integrum* means restoration to the previous condition/original status.

Therefore, following the Methodology for Calculating the Size of Compensation for Environmental Damage,[24] 2.7 million EUR (9,165,371,016 LTL) were specified as the amount for the residual environmental damage remaining after decontamination and neutralization operations implemented by the petroleum company to restore the environment.[25]

According to the law, environmental restoration measures shall be included in calculating the final amount of the compensation. It needs to be clarified that we are not talking about the financial value of the neutralization or decontamination works but the extent to which the state of the environment has actually been restored from the moment of the detection of damage.

Calculation formula: 42,429 EUR (146,500 LTL) (for every ton of residual pollution in soil) × 1289 [rate based on the consumer price index (CPI)] × 32,357 (total amount of residual oil in tons) × 1, 5 (index for a category of a contaminated area).

The formula shows that calculation is based on various factors such as the nature of the pollutant, the type of contaminated land, the consumer price index (CPI), calculated annually by the Statistics Department. The compensation rate is also determined by various coefficients. The more environmentally rich the territory is, the higher the coefficient value is applied. However, the CPI does not reflect the value of environmental features but rather reflects a general shift in prices, i.e. the inflation rate. This means that nature compensatory methodologies are inherently dependent on an economic indicator.

While most of the diesel contamination could be removed in the presented case study, no information is provided on the total value of the lost natural resources such as destroyed ecosystem services. Although a range of indicators (e.g. category of a territory, type of a pollutant) are taken into account, it is doubtful if the compensation based on the quantities of diesel removed from the affected location could be sufficient to define the value of the total damage to nature. On the other hand, this civil case constitutes an important contribution to environmental legislation by clarifying the relationship between restoration of the environment to its original state (*restitutio in integrum*) as well as equivalent monetary compensation within the national compensation system.

[24]Order of the Minister of the Environment of the Republic of Lithuania, September 9, 2002, No. 471 "On the Approval of the Methodology for Calculating the Size of Compensation for Environmental Damage" (*Official Gazette, 2002-09-25, No. 93-4026, as amended—TAR, 2014-09-01, No. 11566.*) The methodology is only applicable for the contamination of ambient air, water, soil and/or the soil substrate. The established formula are intended to assess a damage to the environment (its components), which occurs in the future.

[25]Ibid, 23.

2.3 Conclusions on the Potentials of Biodiversity Offsets and NNL in Lithuania

1. The EIA Directive, the SEA Directive, the Birds and Habitats Directives as well as the ELD emphasize NNL to varying degrees in addressing economic and social impacts. Here the Habitats Directive is the most relevant legislation by establishing the impact mitigation hierarchy and compensatory framework, both of which include offset mechanisms. The conventional and strict application of this highly comprehensive legal instrument is entirely directed towards environmental protection and NNL purposes. It is doubtful whether this unique instrument could be replaced by another legal framework without triggering instability in its satellite conservation mechanisms in national systems.
2. Although NNL is not explicitly established by Lithuanian or EU law, relevant principles for balancing the natural environment, its assets and ecosystems stem from diverse EU requirements. Currently the only mandatory requirements for compensating unavoidable residual impacts on the environment remain issues from the Habitats Directive. As a result, the goal of halting the loss of biodiversity and ecosystem services in Europe remains subject to the functionality of national systems and the readiness to establish an all-inclusive NNL system at EU level. Governments should recognize the added value of a new NNL approach. First, they have to be assured that the offsetting model would not harm existing environmental protection and economic development mechanisms. This could be achieved by integrating particular NNL goals into the details of general development plans or land-use planning documents.
3. Whereas transposition of the Habitats Directive aims to establish and maintain a coherent European ecological network of protected species and habitats, the EIA seeks to balance environmental, political and other public interests in developing economic activities across the whole country. Procedurally, SEA and EIA do not establish measures towards NNL as they do not provide a mandatory legal requirement to compensate for residual impacts on biodiversity and ecosystem services. The Lithuanian EIA regulation illustrates the option for the competent authorities to reject a proposed economic activity at a selected location if likely adverse effects on the environment, including biodiversity, may arise. As a result, the EIA cannot be viewed merely as a procedural instrument without any material consequences.
4. Environmental compensation incorporates certain forms of civil liability requiring restoration of the environment to its original state (*restitutio in integrum*) and compensation for all costs. The more vulnerable and irreplaceable the affected biodiversity, the more strict the implementation of *like-for-like*. However, in cases where biodiversity and ecosystems are vulnerable or irreplaceable, the priority should be to establish strictly protected territories rather than seeking compensation, especially, monetary. However, as the Lithuanian framework of environmental liability illustrates, the relationship between primary and equivalent monetary compensation depends on the national system. The resulting

broad range of compensatory principles between Member States hinders the creation of a uniform environmental compensation mechanism at EU level.

References

Communication from the Commission to the European Parliament, the Council, the Economic and Social Committee and the Committee of the Regions "Our Life Insurance, Our Natural Capital: An EU Biodiversity Strategy to 2020" (COM/2011/0244 final)
Council Directive 92/43/EEC of 21 May 1992 on the conservation of natural habitats and of wild fauna and flora (OJ L 206, 22.7.1992, p 7)
Decision of Kaunas Regional Department of Environmental Protection (2012) No. (PAV)-D2-1280, "On the Proposed Economic Activity in the IIIrd Gravel Pool of Kalnėnai in Jurbarkas District", 29 May 2012
Directive 2014/52/EU of the European Parliament and of the Council of 16 April 2014 amending Directive 2011/92/EU on the assessment of the effects of certain public and private projects on the environment (OJ L 124, 25.4.2014, p 1–18)
Directive 2004/35/CE of the European Parliament and of the Council of 21 April 2004 on environmental liability with regard to the prevention and remedying of environmental damage (OJ L 143, 30.4.2004, p 56–75)
Directive 2001/42/EC of the European Parliament and of the Council of 27 June 2001 on the assessment of the effects of certain plans and programmes on the environment (OJ L 197, 21.7.2001, p 30–37)
Gumbis J, Bacianskaitė V (2014) Lithuania: National individualism in the EU. Comp Public Law Rev 16:165–191
Law on the Environmental Protection of the Republic of Lithuania No. I-2223, Lietuvos aidas, 1992-01-30, Nr. 20-0
Law on the Environmental Impact Assessment of Proposed Economic Activity of the Republic of Lithuania No. I-1495, Official Gazette, 2013-07-16, No. 76-3835
Law of Protected Areas of the Republic of Lithuania No. XII-1305, TAR, 2014-11-13, No. 2014-16784
Law on Forests of the Republic of Lithuania No. XI-1448, Official Gazette, 2011 No. 74-3548
Manual for Environmental Impact Assessment in Lithuania/Raulinaitis M, Auglys V, Bučiūnaitė I et al. (2001) Ministry of the environment of the republic of Lithuania, Finnish Environment Institute, Vilnius
Official website of the Ministry of Environment of the Republic of Lithuania. www.am.lt
Official website of the State Service for Protected Areas of the Republic of Lithuania. www.vstt.lt
Order of the Minister of the Environment of the Republic of Lithuania (2005), No. D1-636, "Approval of Regulations on Preparation of the Environmental Impact Assessment Programme and Report", 23 Dec 2005, Official Gazette, 2010-07-27, No. 89-4729
Order of the Minister of Environment of the Republic of Lithuania (2002a) No. D1-255, "On the Approval of Procedure on the Determination the Significance of the Implementation Impact of Plans or Programmes and Proposed Economic Activities on Existing or Potential Natura 2000 Sites", 22 May 2006, Official Gazette, 2006-05-31, No. 61-2214
Order of the Minister of Environment of the Republic of Lithuania (2002b) No. 471 "On the Approval of the Methodology for Calculating the Size for Compensating the Environmental Damage", 9 Sep 2002, Official Gazette, 2002-09-25, No. 93-4026, TAR, 2014-09-01, No. 11566
Ragulskytė-Markovienė R (2009) Planuojamos ūkinės veiklos poveikio aplinkai vertinimo teisinio reguliavimo problemos. TEISĖ 72

Resolution of the Government of the Republic of Lithuania (2011) No. 1311 "Procedure on Forest Land Conversion and Compensation", Official Gazette, 2011, No. 120-5657
Resolution of the Government of the Republic of Lithuania (2012) No. 1523, Official Gazette, 2012, No. 146-7536
Ruling of the Constitutional Court of the Republic of Lithuania (2013) "On the Procedure of the Publication of Legal Acts" of 29 Oct 2003, No. 1/02
Ruling of the Supreme Court of Lithuania (2014) Lietuvos Respublikos aplinkos ministerijos Panevėžio regiono aplinkos apsaugos departamentas v. AB "Orlen Lietuva", No. 3K-3-61/2014

Spain

David Álvarez García, Marianne Darbi, and Wolfgang Wende

1 Theory in Spain

1.1 Legal Background in Spain

As has been emphasized throughout this book, EU law provides much of the legal framework that underpins biodiversity offsetting. In the case of Spain, the transposition of EU Directives into national environmental legislation is not complete until regional legislative bodies approve their own law. This is because each Autonomous Community (formal region) in Spain has its own competences in terms of environmental legal jurisdiction. Hence, even when there is a national legal framework, the legal obligations have to be translated and enforced in the various regions.

This has two important implications:

- Legal obligations differ among regions.
- What is mandatory in one region might not be mandatory in another.

D. Á. García (✉)
Ecoacsa Reserva de Biodiversidad, Madrid, Spain
e-mail: davidalvarez@ecoacsa.com

M. Darbi
Helmholtz Centre for Environmental Research, Leipzig, Germany
e-mail: marianne.darbi@ufz.de

W. Wende
Leibniz Institute of Ecological Urban and Regional Development and TU Dresden, Dresden, Germany

National University of Singapore, Singapore, Singapore
e-mail: w.wende@ioer.de

W. Wende et al. (eds.), *Biodiversity Offsets*,
https://doi.org/10.1007/978-3-319-72581-9_10

The main legal obligations related to issues of environmental conservation are determined by the EU Habitats Directive[1] and Birds Directive.[2] The Environmental Liability Directive[3] and the Environmental Impact Assessment Directive[4] complete the legal framework with respect to protection against threats and impacts.

Against this background, in the following we analyze the legal framework for biodiversity offsetting or compensation measures from the national perspective. The chapter closes with some case studies to reflect these considerations in practical terms.

Further policies and requirements must be considered beyond this legal mandatory framework. Probably the principal of these is the European 2020 Biodiversity Strategy,[5] which sets the stage for future legal frameworks. This means that all work done in order to meet the 2020 Strategy can help to anticipate potential legal obligations that may subsequently arise. In the Strategy the EU points out the importance of halting biodiversity loss by concluding CBD agreements while emphasizing the importance of biodiversity and its ecosystem services for European wellbeing. With respect to compensation measures, the 2020 Strategy highlights the importance of restoring ecosystems whenever possible, introducing the goal of *"No net loss of biodiversity and ecosystem services"*. This consideration is included in Target 2, Action 7b of the EU biodiversity strategy. One way of achieving this goal is to implement new compensation schemes for biodiversity losses.

At the national level in Spain, the *"Strategic Plan for Natural Heritage and Biodiversity"*[6] for the period 2011–2017 states that it is necessary to adopt measures to halt biodiversity loss from development projects and construction work that impact on the land.

One of these measures, also addressed in the European strategy, is to improve environmental compensation (offsets) for impacts on protected habitats or species. This legally binding Strategic Plan proposes the study and piloting of "*Conservation Banking*" as a way to improve compensation measures.

For this reason, the new Spanish law on environmental impact assessment[7] introduces conservation banking. This national law, which was transposed[8] into all regional frameworks by the end of 2014, was the first to propose conservation

---

[1]Council Directive 92/43/CEE of May 21, 1992 on the conservation of natural habitats and of wild fauna and flora.

[2]Directive 2009/147/EC of the European Parliament and of the Council of November 30, 2009 on the conservation of wild birds.

[3]Directive 2004/35/EC of the European Parliament and of the Council of April 21, 2004 on environmental liability.

[4]Directive 2014/52/EU of the European Parliament and of the Council.

[5]COM/2011/0244 final version "Our life insurance, our natural capital: an EU biodiversity strategy to 2020".

[6]Spanish Royal Decree 1274/2011 of September 16, 2011.

[7]Law 21/2013 of December 9 on Environmental Assessment.

[8]As mentioned at the beginning of this chapter, legislation at national level establishes the basis for environmental legislation, which can be further underpinned and strengthened in each region.

banking as a tool to improve environmental compensation for projects categorized under environmental impact assessment and the liability framework. The Spanish law 21/2013 also includes the novel concept of "conservation credits", providing the option to use these conservation credits as *"compensatory or complementary measures envisaged in the legislation on environmental assessment, environmental liability or natural heritage and biodiversity, in order for the negative effects on a natural value to be balanced by generated positive effects with the same or a similar natural value, in the same or a different place."*

Summing up, we can say that, alongside European directives, there are already various Spanish laws that provide for compensation measures (not necessarily offsetting) in the case of environmental impacts. These are the laws on:

- Natural Heritage and Biodiversity[9]
- Marine Environment Protection[10]
- Environmental Liability[11]
- Environmental Assessment[12]

1.1.1 Compensatory Measures in Spain

To fully understand the legal framework and the practical operation of these rules in the project approval process, it is important to note that the Spanish language does not have separate terms for "compensation measures" and "offsets". The term offset, which implies compensation measures aimed at no net loss or net gain of biodiversity, has no direct translation. While this may appear a trivial issue, it is nevertheless one of the factors hindering the progress of biodiversity offsetting in Spain.

In any case, while ecological compensation (*like for like*) has recently become a common practice in restoration, offsetting is not universally applied to remedy the negative environmental impacts of a project. Thus, the mitigation hierarchy is not always well implemented. Finally, for various reasons most projects only deliver a variety of single compensation measures. Frequently these are not in the form of natural compensation but instead rely on financial compensation or some other non-ecological restoration measures.[13]

---

[9]Law 42/2007 of December 13 on Natural Heritage and Biodiversity, and Law 33/2015 of September 21 on Natural Heritage and Biodiversity. While Law 33/2015 partially modifies law 42/2007, it is important to note that both laws coexist.

[10]Law 41/2010 of December 29 on Marine Environment Protection.

[11]Law 26/2007 of October 23 on Environmental Liability.

[12]Law 21/2013 of December 9 on Environmental Assessment.

[13]Carrasco García et al. (2013).

1.1.2 Natural Heritage and Biodiversity

The Spanish Law on Natural Heritage and Biodiversity (law 42/2007),[14] as well as the amended version of the same law (law 33/2015[15]), make reference to the legal obligations under the Habitats Directive (92/43/EEC) supplemented by the recommendations of the Council of Europe and the Convention on Biological Diversity. For the first time in Spanish legislation, this law also provides for incentives for private initiatives on nature conservation. Land stewardship is one of these examples.

The Law on Natural Heritage and Biodiversity establishes the basic legal framework for the conservation, sustainable use, improvement and restoration of natural heritage and biodiversity in Spain, regardless of property ownership or the legal status of land. The law therefore defines the government's obligation towards conservation and management.

This law also includes catalogues of habitats and species subject to special protection. Regarding habitats, this protection status takes various forms including Natura 2000 areas (derived from the transposition of the Habitats Directive) and national nature protection areas.

Regarding flora and fauna, protection is realized through the recognition of "species of community interest." Annexes I and II of the Spanish law 42/2007 list those species subject to protection. Species may deserve protection due to their uniqueness, rarity, scientific, cultural or ecological value or degree of threat.

A number of implicit prohibitions serve to protect the species included in Spanish law 42/2007 (Annexes I and II). Beyond them, the law also regulates a general scheme and a special protection regime (Art. 52.3, Law 42/2007) and specified plants and animals are included in the list of species under a special protection regime (Art. 54, Law 42/2007).

Under certain circumstances, the law permits activities that may negatively affect these species or habitats. Thus, the Natural Heritage Law (42/2007) provides exceptions that allow and regulate the destruction and alteration of habitats in exceptional and justified cases. In these cases, compensation and reparation measures are required under Spanish law 33/2015 (Art. 46 and Art 58).

This law opens the door for the use of compensation and/or offset measures in the case of damage. These forms of compensation are defined in Article 3.24 (Law 42/2007) as *"specific measures included in a plan or project which are intended to compensate, as closely as possible, its negative impact on the species or habitats affected."* Concerning Natura 2000 sites, Article 46 (Law 33/2015) regulates the necessary measures in case of impacts; in particular, paragraph 5 (Art. 46.5 Law 33/2015) determines compensatory measures with regard to projects that affect the Natura 2000 network but are justified by an overriding public interest of the highest order.[16]

[14]Law 42/2007 of December 13 on Natural Heritage and Biodiversity.

[15]Law 33/2015 of September 21 on Natural Heritage and Biodiversity.

[16]The law itself determines that a project must meet particular requirements in order to be declared as being of public interest.

However, habitats and species of community interest are not only found in Natura 2000 sites. These are present elsewhere, for example in other nationally or locally protected areas such as nature reserves or national parks, where conditions also need to be assessed. Article 80 of the Spanish law on Natural Heritage and Biodiversity (law 33/2015) specifies that interventions resulting in the deterioration or alteration of priority habitats will result in sanctions. Clearly, the compensation measures for these cases must be investigated.

It is important to point out that the most recent amendment of Law 33/2015 introduced the principle of avoiding net loss of biodiversity, which is mentioned in the preamble as one of the law's guiding principles. However, this principle is not developed within the law, i.e. no further restrictions are specified in this regard. Thus, according to Article 61.2 (Law 35/2015), the principle is to be implemented by the competent administrations in each case. The article also adds new exceptions to some prohibitions included in the law to reflect the proper implementation of the Habitats Directive.

1.1.3 Marine Environment Protection

Spain's Marine Environment Protection law sets out the obligations with respect to biodiversity conservation and establishes marine protected areas.

In its first article, the law mandates the protection of the marine environment, the prevention of deterioration and recovery of marine ecosystems, which also means compensating/offsetting impacts. Anyway, this law does not clarify implicitly nor explicitly whether or not compensation is required.

1.1.4 Environmental Liability

Spain's Law 26/2007 on Environmental Liability was passed in 2007. It fully transposes the European Directive requirements and measures into Spain's legislation. Regarding ecological compensation (*like for like*), the adoption of this law marked a turning point in the country's legal framework regarding reparations for incidental environmental impacts. Under the law, ecological reparation became compulsory in the case of incidental environmental impacts regardless of the cost. This was an entirely new principle in Spain's environmental legislation.

However, while this law provides considerable backing for the compensation and remediation of environmental impacts, it has only been invoked in several instances over the past decade.

The main features of the Environmental Liability law are:

- Unlimited liability for reparation. Affected resources and services must be restored to their original state regardless of the cost.

- Clear definition of responsibilities: in a more general way (activities included in Annex III) and in an individual way (activities not included in Annex III), as well as mandatory prevention measures.
- Protection of soil, water, wildlife and protected habitats, seashores and estuaries (as subjects of protection due to the Environmental Liability law) from harm and threats of harm. Through this law, it is mandatory to restore any harm and threat of harm regardless of restoration costs.

The law is developed further by the Spanish Royal Decree 2090/2008.[17]This regulation defines three kinds of reparation in the case of incidental damage. The first of these is:

- Primary reparation, i.e. returning natural resources and the ecosystem services to their baseline.

Although primary reparation must always be the first aim, it may not be feasible in some cases. Clearly, full restoration of natural resources and ecosystem services will not be possible after some major impacts. Or full restoration may only be achieved after a substantial period of time. In these cases, complementary reparation or compensatory reoparation are specified:

- Complementary reparation: Any additional corrective measure implemented with regard to natural resources or natural resource services to compensate for the fact that primary reparation does not result in full ecological functioning.
- Compensatory reparation: Compensation for interim losses of natural resources or services from the date on which the damage occurred until the time at which primary reparation has achieved its full effect. This is the primary goal of conservation banks.

In order to meet the legal requirements, these complementary reparation or compensatory reparation measures could be realized by a conservation banking system. However, the scarcity of practical cases brought under the law as well as the lack of a regulated model to calculate losses and gains makes it difficult to implement a conservation banking system as a way to meet the legal requirements set by this act (law 26/2007).

1.1.5 Environmental Impact Assessment

The environmental impact assessment procedures provide mechanisms to propose the most appropriate way to implement a project while minimizing the impact on the environment.[18]

[17]Royal Decree 2090/2008, which brings into force the partial Rule of Law 26/2007 of Environmental Liability.

[18]Garmendia et al. (2005).

Regarding environmental assessment, European Union law is shaped by two directives. These are Directive 2001/42/EC of 27 June 2001 on the assessment of the effects of certain plans and programmes on the environment, and Directive 2011/92/EU of 13 December 2011 on the assessment of the effects of certain public and private projects on the environment.

These directives are transposed into Spanish legislation via the Spanish Environmental Assessment Law 21/2013. This new legal framework gathers in one legal text the regulation for plans and programmes as well as individual projects. Law 21/2013 also provides new tools and aspects that were not included in the previous version of this law,[19] including conservation banking, compensatory measures beyond the Natura 2000 network and a new definition for interim impacts. This new perspective opens a pathway for compensation measures, offsetting and similar tools, but, they are not resulting in offsetting practice yet.

However, Act 21/2013 was not the first law to introduce ecological compensatory measures to Spain. Before 2013 we can also point to instances of ecological compensatory measures. The first example of ecological compensatory measures under environmental assessment legislation can be found in the Itoiz dam project of 1994.[20] Here the Environmental Impact Declaration included ecological compensatory measures, specifically off-site measures in a different river basin from that affected. Previously, compensation measures in Spain had only been related to socio-economic measures.

The fundamental step in the application of compensatory measures in Spain was the incorporation of Article 6.4 of the EU Habitats Directive (92/43/EEC) into Spanish law. This incorporation included the requirement for compensatory measures in certain circumstances. Hence, broadly speaking and without reference to the regulations on environmental impact assessment of the different autonomous communities, the three main requirements for compensatory measures can be specified as:

- Compensatory measures required by Article 6.4 (EU Habitats Directive) associated with significant[21] impacts on the Natura 2000 network and sites.
- Compensatory measures for important but non-significant impacts on the Natura 2000 network (location, size of the project . . .).
- Compensatory measures outside Natura 2000, mainly driven by impacts on habitats of community interest and/or habitats or species of interest.

[19] Law 9/2006 on plans and programmes and Royal Decree 1/2008 on environmental impact assessment.

[20] BOE (Official State Gazette) 09/06/1990.

[21] Under Spanish legislation on Environmental Assessments, the significance of an impact is defined as: "Permanent or long term alteration of natural values and in the case of Natura 2000 sites, alteration which also affects the elements motivating its designation as a Natura 2000 site or its conservation goals".

However, there has been no wide application of this legal framework related to compensation measures. Until now it has been limited to compensatory measures in application of Article 6.4 of the EU Habitats Directive.

The new Spanish regulations in force since 2013 in Environmental assessment aim to address these gaps with respect to compensation measures and offsetting by widening their scope and, more importantly, clarifying the rules for the definition of compensation measures. Here it is useful to point out the main features of the law:

- If the impact assessment declaration includes ecological compensation measures, these cannot be replaced by any other non-ecological compensation measures.
- The law requires interim impacts to be calculated and quantified. However, the law does not specify how to estimate the impacts or which methodology to apply. The definition of significant damage must be more clearly defined.
- The law clarifies the circumstances under which compensatory measures are required.
- It includes for the first time *conservation banking* as an optional tool for ecological compensation.

As mentioned earlier, the Spanish Environmental Assessment Law 21/2013 specifies nature conservation banking for the first time. Specifically, the additional provisions (VIII) of the law propose conservation banking as a tool for ecological compensation. Although the law defines the basic features of conservation banking, it is still necessary (and the law also alludes to this) to establish future guidelines and rules for the creation and use of conservation banks to offset environmental impacts.

1.2 Methodological Background in Spain

1.2.1 Mitigation Hierarchy in Spain

As introduced in the discussion on the legal background, there exist several Spanish laws that transpose European environmental directives. The country's legal context for the mitigation hierarchy is formed by three separate laws on environmental assessment,[22] natural heritage[23] and environmental liability.[24] All of these make reference to the mitigation hierarchy, i.e. the avoidance, minimization, reparation and compensation of environmental impacts.

Environmental assessment procedures were introduced to Spain in 1986 with the transposition of the 1985 European directive[25] on the "Assessment of the effects on

[22]Law 21/2013 of December 9 on Environmental Assessment.

[23]Law 33/2015 of September 21 on Natural Heritage and Biodiversity.

[24]Law 26/2007 of October 23 on Environmental Liability.

[25]Council Directive 85/337/EEC of 27 June 1985 on the assessment of the effects of certain public and private projects on the environment.

the environment of certain public and private projects". Environmental impact assessment has subsequently undergone a remarkable evolution. Although the principle of a mitigation hierarchy was included in environmental assessment legislation prior to the new Spanish Environmental Assessment Law 21/2013, this law reinforced the principle so that the objective of offsetting impacts could be fulfilled by natural/in-kind compensation.

Under the Environmental Liability Law (26/2007),[26] the principle of prevention and avoidance is applied to many activities not previously subject to environmental assessment procedures. Environmental liability law is now evolved in terms of procedural criteria of minimizing risks, replacement and compensation from impacts to economic activities that are a threat to the environment. Although many procedures have been developed under this law, it has yet to be applied to more than a handful of economic sectors. Chapter III of Law 26/2007 determines the framework for the "Prevention, avoidance and reparation of environmental damage", which aims to achieve a no net loss of environmental resources.

Both legislative frameworks—Environmental Assessment (21/2013) and Environmental Liability (26/2007)—have encouraged many organizations to include the concept of no net loss in their business management, especially in activities that are directly linked to the environment. However, much still needs to be done to introduce this concept into business management beyond legal requirements.

1.2.2 Metrics in Spain

The Environmental Liability Law has fostered the adoption of metrics to measure environmental impacts. Specifically, the law establishes a set of procedures to calculate and measure environmental impacts.

Regarding environmental assessment, the new Spanish Environmental Assessment Law (Law 21/2013) requires that the main impacts be measured both qualitatively and quantitatively within the environmental assessment procedure.

It established the need to *"identify, quantify, and determine the extent of damage, as well as its intensity, time scale and significance"* (Art. 8, Law 21/2013).

The Environmental Assessment Law also provides a set of criteria to quantify the impacts identified as significant (Annex VI, Law 21/2013). It specified the need to quantify how the resource is impacted according to various environmental indicators, such as:

- Condition of the resource,
- Quantitative ecological status of the affected resource,
- Physical integrity of the affected resource,
- Quality of the affected resource,
- Risks to human health or the environment associated with the affected resource,
- Its structure and function.

[26]Law 26/2007 of October 23 on Environmental Liability.

Although these indicators help to quantify impacts, there is no specific metric available to estimate the impact precisely. The Environmental Assessment Law requires that impacts be evaluated *"whenever possible, based on the quantification, using for this purpose the methodologies provided in rules or technical studies that are usually applied"*. Hence, the Spanish Environmental Assessment Law does not directly specify methodologies to measure impacts but merely determines which aspects should be taken into account in the calculation.

The Spanish Environmental Assessment Law includes conservation banking as a tool to compensate for environmental impacts. Currently, however, the law neither defines specific metrics for conservation banking nor ways to implement them. These aspects will be developed by further legislation. According to Additional Provision No. 8 and Article 7 of Law 21/2013, the organization, functioning and technical criteria of nature conservation banking will be legally established in the future. Thus, we can expect the issuance of specific guidelines on conservation banking in the years to come, detailing the use of conservation credits as a compensatory measure. However, it is currently unclear upon which legal authority these guidelines will rest.

1.2.3 Offset Quality Assurance in Spain

As mentioned in the opening section, there is no equivalent term in Spanish for "offsets". After the implementation of the environmental assessment procedure under the Royal Decree 1302/1986, the application of compensatory measures in Spain has developed rather unevenly, especially as regards forestry and agricultural landscapes.[27] Compensatory measures were not officially required until the release of Directive 92/43/EEC in May 21, 1992. Article 6.4 of this directive establishes mandatory compensatory measures for impacts on Natura 2000 sites under certain conditions. Following the transposition of this directive into Spanish law in 2000, compensatory measures began to be applied in a standard fashion. These measures include requirements to ensure that an impacted site is completely "restored" to its state before the intervention.

The quality of compensatory measures in Spain is highly variable, depending on whether they are designed to comply with Article 6.4 of the 92/43/EC Directive for compensatory measures in the Natura 2000 network or for other environmental law. The former measures have very strict requirements to restore the coherence of the Natura 2000 network. The latter measures, usually called "accompanying measures", are defined by the Spanish administration as compensation for any impacts not directly related to Natura 2000. Although implemented in a large proportion of projects subject to environmental impact assessment, these accompanying measures do not have the same quality as compensatory measures pursuant to Article 6.4. In most cases they are intended to improve rather than restore damaged resources or are realized in timeframes unrelated to the damaged resource.

[27] Carrasco García et al. (2013).

Therefore, there is no guarantee that the quality of accompanying compensatory measures in Spain achieves that required in cases where the coherence of the Natura 2000 network is affected. As a means of enforcing higher offset quality, conservation banking is established under the Environmental Assessment Law.

1.2.4 Methodology of Habitat Banking/Compensation Pools/Eco Accounts in Spain

The Spanish Environmental Assessment Law (21/2013) articulates the framework for the use of conservation banking to develop compensatory measures for impacts or incidental impacts under the Law on Environmental Liability (26/2007).

Nevertheless, this framework is only presented in a very simplified way. The law merely stipulates the objectives of nature conservation banking and the main characteristics that conservation banks must meet in order to be used as compensation tools. The requirements to meet the bank's nature conservation objectives are included in an Additional Provision No. 8 (Law 21/2013), which lists seven legal requirements for this purpose.

While conservation banks are established under the Spanish Environmental Assessment Law (21/2013), conservation credits generated in conservation banks can be used as "compensatory measures under the Environmental Assessment (21/2013), Environmental Liability (26/2007) or Natural Heritage and Biodiversity Laws (42/2007 and 33/2015)", opening the possibility for their use beyond compensation in the scope of environmental assessment.

The conservation banking model has been approved by means of the Environmental Assessment Law. In particular, the various articles of the law specify the following aspects:

1. Nature conservation banks are defined as a set of environmental conservation titles or credits, granted by the Ministry of Agriculture, Food and Environment and the autonomous communities, which represent natural values specifically created or improved. This definition is given in the Spanish Environmental Assessment Law and includes guidance on the banks' operation in regard to conservation such as the titles granted (property rights) and additionality by defining these titles as *"natural values specifically created or improved"*.
2. A bank is created by an administrative decision. This decision identifies land register data as well as the number of conservation credits and technical criteria that should be adopted for bank creation and future conservation.
3. Landowners affected by banks must conserve over the long term natural values created or improved. The law also establishes that these sites must only be used for purposes that are compatible with the aforementioned natural values. There is a requirement to register future conservation banks with the Real Estate Property Registry. This registry includes restrictions on land use according to the goals and management plan for the conservation banks. Restrictions mentioned in this article are established in order to ensure the preservation of natural values.

4. The Spanish Environmental Assessment Law permits the sale of credits on the open market. It also mandates that such a sale be recorded in a national register of conservation banks covering the whole of Spain. If registers are later created in individual regions, these should be integrated into the national register.
5. The Environmental Assessment Act sets out a regime of sanctions for any breach of conservation banking rules.

A sound methodological framework is needed to further develop conservation banking in Spain. This methodological framework will be established by the issuance of a Royal Decree, in which the following key aspects will be developed:

1. Conservation objectives: While the Environmental Assessment Law does not identify which natural values can be created, restored or enhanced in conservation banks, the objectives for conservation banking in Spain are likely to be related to the compensatory measures required by the Environmental Assessment Law. These measures are almost always linked to significant impacts on Natura 2000 sites or to other impacts on priority habitats or species of community interest that must be compensated in the opinion of the competent authorities.
2. The land which conservation banks can manage has not yet been specified in the Environmental Assessment Law. It is important that a conservation banking regulation be established, including a set of eligibility criteria for developers of conservation banking. This will also allow the development of national conservation strategies as well as a series of specific conservation objectives.
3. Regarding additionality, although conservation banks are defined by the Spanish Environmental Assessment Law as a means to create and improve natural values, no conservation banking requirements are in place for a proper design of conservation measures.
4. The methodologies to be applied for quantifying conservation credits and for equivalences between credits and debits have still to be defined.
5. Long-term safeguards are key to the proper functioning of conservation banking. These conservation safeguards are not fully reflected in the Environmental Assessment Law (21/2013).

Clearly, a sound methodological framework is needed to further develop conservation banking in Spain. This methodological framework will be established by the issuance of a Royal Decree, in which all of these key aspects will be developed.

2 Practice in Spain: Designing an Environmental Impact Compensation Scheme Based on Conservation Banking

Background

In 2011 the Strategic Plan for Natural Heritage and Biodiversity was enacted as Law no. 42/2007.[28] The plan takes account of the environmental impact assessment in relation to biodiversity. In this way, the Strategic Plan for Natural Heritage and Biodiversity[29] states that *"the full consideration of biodiversity in processes of environmental assessment is essential to insure its correct protection and conservation"*.

With the aim of achieving biodiversity objectives, it is necessary to take account of the environmental impact assessment and to apply methods of environmental impact compensation while obeying the mitigation hierarchy.

However, the legal text mentions,[30] *"in reality, the compensation measures (...) which respond to Article 45 of Law 42/2007,* **are not always suitably designed and do not conform to the principle of compensation for damage caused to habitats or species.** *For these reasons, they are viewed as insufficient for the intended purpose. Ideally, these measures should be designed as to actively involve the managerial organs of Natura 2000, in particular, as well as to encourage the conservation of biodiversity."*

Furthermore, at a national level, the document recognises that *"deficiencies are detected in the monitoring of the application and efficacy of the preventative, corrective and compensatory measures"* such as the *"inadequate verification of compliance with the established environmental conditions nor for the carrying out of the necessary on-site checks"*.

Hence, such deficiencies in the design and monitoring of compensation must be remedied so that measures can achieve the objectives of the environmental impact assessment and to conserve natural heritage and biodiversity, especially in sensitive areas as well as Natura 2000 sites.

Due to this situation, which has contributed to the deterioration of biodiversity not only in Natura 2000 sites but also in locations outside of this network, the plan considered it essential to *"reinforce the capability of the environmental authority in its monitoring and follow-up work."*

Specifically, Action Point 8.1.7 of the plan mentions the need *"to study and regulate, if necessary, the implementation of Conservation Banking."* This action is consistent with Goal 8, namely *"to mobilize financial resources from all sources to achieve the objectives of the conservation of biodiversity"* and Objective 8.1, which is *"to ensure the adequate financing of biodiversity conservation policies."*

---

[28]Natural Heritage and Biodiversity Law 42/2007.

[29]Royal Decree 1274/2011, dated 16 September, which adopted the Strategic Plan for Natural Heritage and Biodiversity 2011–2017.

[30]Law 21/2013 of December 9 on Environmental Assessment.

The National Framework
An immediate result of a ministerial working group from the Ministry of Environment (originally MARM[31] later MAGRAMA[32]) was the inclusion of the Strategic Plan for Natural Heritage and Biodiversity under Action Point 8.1.7.

The first real legislative outcome in relation to the valuation of environmental impact and biodiversity was enshrined in Law No. 21/2013, which for the first time recognized conservation banking at a national level in its Eighth Additional Disposition.

The law provides:

1. The first definition of Conservation Banking that serves to establish a new development framework: *"Conservation Banking is a market in which the credits from actions (...) that enhance and significantly improve biodiversity outcomes can be purchased to offset debits from environmental damage."*
2. A biodiversity credits framework for Conservation Banking: *"Biodiversity credits could constitute compensation or complementary offsets provided for in environmental assessment legislation, environmental responsibility or regarding natural heritage and biodiversity"*
3. An additional regulation that specifies *"general, organizational, functional and technical criteria of Conservation Banking, which will develop according to the defined rules (...)".*

Regulatory Procedures
Law 21/2013 initiated the environmental impact assessment process for the drafting of the Conservation Banking regulations. This process culminated in the first draft of the regulation, which recognized all core elements needed to establish a system of Conservation Banking.

The text included important innovative and legal provisions concerning the environment, marking a major step in the development of instruments for the conservation of natural resources. One basic innovation was the possibility of registering environmental values, thereby generating conservation credits in the Property Register. This registration attached a "condition" to the registered land in the form of a marginal note in the registry document affecting any sale or dispossession of land subject to conservation banking. In this way it became possible to retain obligations regarding nature conservation that were created and registered before any change in asset ownership.

A number of Spain's autonomous communities were involved in the consultation process to draft the Regulation as well as interested stakeholders, i.e. private property developers, environmental NGOs, businesses, property owners, insurance companies and environmental consulting agencies.

[31]Ministerio de Medio Ambiente, y Medio Rural y Marino (The Ministry of the Environment, Rural and Marine Affairs).

[32]Ministerio de Agricultura, Alimentaci6n y Medio Ambiente (The Ministry of Agriculture, Food and Environment).

Although the draft text of the regulation addressed most of the demands of the different stakeholders, it was not adopted due to a general lack of consensus. Some specific problems were[33]:

- A lack of agreement with the main environmental organizations. By definition, Conservation Banking is the last step in the mitigation hierarchy, in so far as it refers to residual impact compensation. According to environmental organizations, the Spanish legal draft did not correctly define the framework of the residual impacts or the mechanisms for their quantification, nor did it establish criteria by which these impacts should be compensated. Without a clearly defined framework, it is impossible to establish compensation mechanisms. This is a problem particularly for locations outside of Natura 2000, which are not subject to the existing requirements of the Habitats Directive (Art 6).
- The draft regulation was nevertheless excessively protectionist in the sense that once it had finally been approved and put into action, there was the possibility that it would not be used; specifically, the requirements for the creation of Conservation Banking, and ultimately for compensation through generated credits, would be greater than those laid down in any of the usual project-to-project compensation measures *(case by case, like for like, on site).*
- The use of Conservation Banking within Natura 2000 created disputes.
- A lack of participation in the development of the model. While it is true that the public became involved after publication of the first draft, there was little prior consultation regarding the preliminary concepts for the model of Conservation Banking presented in this draft.
- Associated aspects of the liability scheme were not completely considered.

2.1 Case Study: Murcia

Alongside the analysis process and studies on Conservation Banking undertaken on behalf of the Spanish Government, other Autonomous Communities analyzed the possibility of developing a practical Conservation Banking model by virtue of the jurisdiction in environmental management accorded them by the constitution. It is worth recalling that in Spain, jurisdiction in environment matters resides within the various autonomous communities. The Ministry of the Environment is only involved in the case of projects for which the substantive body is the General Administration of the State.

In this way, in 2012 the region of Murcia (an Autonomous Community) included a proposal for the development of "Conservation Banking" in its "Draft Law of Conservation and Biodiversity".

Although the draft law was not enacted and, consequently, the proposed model of Conservation Banking was not established, the inclusion of these mechanisms in the

[33]Alvarez et al. "Custodia del territorio y bancos de conservación." (Land Stewardship and Conservation Banking) CONAMA 2014.

draft legal text is a good example of how autonomous administrations, in this case in Murcia, can develop this type of instrument.

This case involved a model oriented around micro-impact compensation involving the creation of an "*Administrative Register*" of regional character different to the model proposed by the Ministry. As an administrative action, this would meet objectives that could not be achieved by other standard procedures, thereby securing sources of funding for the conservation of biodiversity.

Under the proposed Administrative Register, land located in the Region of Murcia should be voluntarily registered for the implementation of biodiversity conservation actions.

This register additionally included:

- Technical tasks necessary for the valuation of compensation measures;
- Assignment, monitoring and control of conservation projects;
- Supervision of transactions;
- Dispute resolution between stakeholders.

Although the proposed system did not provide a defined metric, the basic design was intended to allow the registration of environmental gains within the environmental accounting system SEEA (System of Environmental-Economic Accounts) developed by the United Nations Statistical Commission, where they would be classified as "intangible environmental assets".

The simultaneous regulatory amendments by MAGRAMA slowed down the development of this innovative model, which is linked not only to impact compensation but also the value-creating actions of environmental accounting.

2.2 Catalonia

The case of the Generality of Catalonia (the official name for the Autonomous Community of Catalonia) is similar to that of Murcia, despite different development processes and results.

Both the Generality of Catalonia and the Region of Murcia demonstrated interest in Conservation Banking once the concept was included in the Strategic Plan of Natural Heritage and Biodiversity.

However, before developing its own model, Catalonia hoped to learn more about the Spanish Ministry of the Environment's proposal to undertake the work by virtue of its territorial jurisdiction and its framework for action.

In this way, Catalonia conducted an early study regarding the development of its own system for Conservation Banking, entitled "Background Study for the Implementation of a System of Conservation Banking in Catalonia".

This project was finalised after several days of public seminars involving different stakeholders, who were given the opportunity to voice their opinion on a possible development model.

The process was directly followed by an analysis of various legal aspects in order to define the scope of the banking tools.

Conservation Banking in Catalonia falls within a No Loss Network, part of biodiversity programme, which intends to carry out other biodiversity protection measures such as green infrastructure on behalf of the Generality.

3 Conclusions from a Spanish Practitioner's Perspective on an EU-Wide No Net Loss and Offset Strategy

Although Spain was one of the early EU member states to develop conservation banking in its legal framework and environmental assessment, there is still a lack of standard rules for the proper creation of conservation banks. Thus, presumably there are no banks in place in Spain yet.

If we consider that the Environmental Assessment Act was passed at the end of 2013, we have to ask why no guidelines have yet been issued on the establishment of conservation banks. There are a number of reasons for this, several of which are no doubt political. One reason, though, is perhaps simply that the Spanish environmental sector is still unprepared for this tool.

The Spanish transposition of the different strategic components of the European Biodiversity Strategy to 2020 will probably help encourage the establishment of such guidelines. However, this will depend on the sociopolitical framework and, moreover, on a change in the application of and requirements for compensatory measures.

Furthermore, the lack of demand for conservation credits from conservation banks is one barrier to a well-functioning market for compensatory measures. Alongside the absence of a system to compensate interim impacts and the lax enforcement of compensatory measures, this missing market is probably the main challenge to be overcome in the successful development of conservation banking in Spain.

At the same time, it should be remembered that various stakeholders (e.g. landowners) are urging the introduction of conservation banking and attempting to solve a range of related problems, including ongoing negative impacts as well as limited income opportunities in rural landscapes.[34]The creation of a market in biodiversity credits has the potential to address these challenges.[35]Funded by the Ministry of the Environment, a Spanish foundation called the *Fundación Biodiversidad*[36] has also provided some recommendations[37] regarding the financing of nature conservation.

[34]Manifesto for conservation banking (http://www.mercadosdemedioambiente.com/actualidad/manifiesto-a-favor-de-los-bancos-de-conservacion-en-espana/).

[35]Alvarez D., González Alcalde I. Bancos de hábitat una solución de futuro. (http://www.mercadosdemedioambiente.com/docs/100_doc_96892140.pdf).

[36]*Fundación Biodiversidad* is an organization belonging to the Ministry of the Environment.

[37]Alvarez D., González Alcalde I. Bancos de conservación de la naturaleza (http://prioridadrednatura2000.es/sites/default/files/lifemap_bancos_de_conservacion.pdf).

References

Alvarez D, González Alcalde I (2014a) Bancos de conservación de la naturaleza. http://prioridadrednatura2000.es/sites/default/files/lifemap_bancos_de_conservacion.pdf
Alvarez D, González Alcalde I (2014b) Bancos de hábitat una solución de futuro. http://www.mercadosdemedioambiente.com/docs/100_doc_96892140.pdf
BOE (Official State Gazette) (1990) 09/06/1990. https://www.boe.es/boe/dias/1990/09/06/pdfs/A26174-26174.pdf
Carrasco García MJ et al (2013) Evolution of the compensation measures in environmental impact assessment procedures in Spain. Ingeniería Civil 172/2013
Council of the European Union (1985) Council Directive 85/337/EEC of 27 June 1985 on the assessment of the effects of certain public and private projects on the environment
Council of the European Union (1992) Council Directive 92/43/CEE of May 21 1992 on the conservation of natural habitats and of wild fauna and flora
European Commission (2011) Our life insurance, our natural capital: an EU biodiversity strategy to 2020
European Parliament, Council of the European Union (2004) Directive 2004/35/EC of the European Parliament and of the Council of April 21 2004 on environmental liability. http://eur-lex.europa.eu/legal-content/EN/TXT/?uri=CELEX:02004L0035-20130718
European Parliament, Council of the European Union (2009) Directive 2009/147/EC of the European Parliament and of the Council of November 30 2009 on the conservation of wild birds. http://eur-lex.europa.eu/legal-content/EN/TXT/?uri=CELEX:32009L0147
European Parliament, Council of the European Union (2014) Directive 2014/52/EU of the European Parliament and of the Council of 16 April 2014
Garmendia A, Salvador A, Crespo C, Garmendia L (2005) Evaluación de Impacto Ambiental. Pearson
Law 9/2006, de 28 de april, about plans and programs and Royal Decree 1/2008, de 11 de enero on environmental impact assessment
Law 33/2015 of September 21 on Natural Heritage and Biodiversity
Manifesto for conservation banking (2015) Manifiesto a favor de los bancos de conservación en España. http://www.mercadosdemedioambiente.com/actualidad/manifiesto-a-favor-de-los-bancos-de-conservacion-en-espana/
Ministry of the environment and rural and marine affairs (2008) Royal Decree 2090/2008, which passes the partial Rule of La 26/2007 of Environmental Liability
Spanish Law 42/2007 of December 13 on Natural Heritage and Biodiversity and Law 33/2015 of September 21 on Natural Heritage and Biodiversity which partially modifies law 42/2007. It is important to point that both laws coexist
Spanish Law 26/2007 of October 23 on Environmental Liability
Spanish Law 41/2010 of December 29 on Marine Environment Protection
Spanish Law 21/2013 of December 9 on Environmental Assessment

The Netherlands

Astrid J.A. van Teeffelen

1 Introduction

The mitigation hierarchy has been part of Dutch spatial planning and environmental laws since 1961 in regard to forests, with extensions to other nature conservation acts since the 1990s. Over the years these laws have changed, along with their interpretation and application. Consequently, the roles and responsibilities of the central and regional governments have also shifted, with responsibilities regarding the enforcement of the mitigation hierarchy being transferred from the national government to the twelve provinces. Since 2017, three nature conservation laws have been merged into a single Nature Conservation Act, with the aim of clarifying and simplifying regulations. In the years to come, the new Nature Conservation Act will be incorporated into an overarching Environment and Planning Act, together with dozens of other pieces of legislation related to spatial planning and the environment. In summary, while Dutch law takes account of no net loss and offset mechanisms, the plurality of designations and regulations applying to sites has proven confusing to both practitioners and regulators. This chapter outlines the main policy mechanisms related to the mitigation hierarchy in the Netherlands, the application thereof, as well as successes and failures as outlined in recent studies. Our particular focus is on biodiversity offsets, which have received much attention in recent years. We discuss the scale of application, the transparency and availability of documentation regarding offset projects, along with the various offsetting mechanisms currently in place (restoration offsets, in-lieu fees). We also briefly touch upon recent developments regarding a habitat banking system and generic metrics to measure no net loss.

A. J. A. van Teeffelen (✉)
Faculty of Science, Environmental Geography Group, Vrije Universiteit, Amsterdam, The Netherlands
e-mail: astrid.van.teeffelen@vu.nl

W. Wende et al. (eds.), *Biodiversity Offsets*,
https://doi.org/10.1007/978-3-319-72581-9_11

2 Key Policies and Developments Concerning No Net Loss and Offsetting

The mitigation hierarchy and offsetting have been part of Dutch conservation policy for over 50 years. The Netherlands has a comprehensive set of no net loss and biodiversity offset requirements, which have been developed for four different types of conservation regimes: forests, Natura 2000 sites, other national nature conservation sites, and protected species. The multitude of rules and regulations, overlapping designations and unclarities regarding roles and responsibilities have, however, hindered the effective practical implementation of policies. While the specifics regarding offsetting impacts for these different regimes were originally established in separate legal frameworks, recently these various laws have been merged into a single Nature Conservation Act (or 'Wet Natuurbescherming'), which came into force on 1 January 2017. At the same time, rules, roles and responsibilities have been clarified. Here we provide an overview of key policies and developments concerning no net loss and offsetting for different types of biodiversity and protection regimes. Table 1 summarises key developments over the last 10 years, while Table 2 gives an overview of the offset rules for the various conservation regimes.

2.1 *Forests*

The Forest Act (or 'Boswet'), which dates back to 1961, generally applies to all woodland sites outside built-up areas. The act contains a provision that the deforestation of areas larger than 100 m^2 or with more than 20 trees within a tree line have to be compensated by afforestation, either on-site or—when on-site afforestation is not possible due, for example, to land use change—off-site. Such afforestation has to take place within 3 years of deforestation. Although the provisions laid down in the act only permit active afforestation through tree planting, in practice natural afforestation is also allowed by most provinces (the regulator).[1] Mandatory compensation of deforested areas is also applicable in cases of deforestation due to fire, storm, diseases or pests. Thinning (up to 40% of tree crown cover) of existing stands and coppice practices are allowed with no requirement of compensation. As the provinces are the regulating authorities, specific provincial rules may be adopted to determine the exact offsetting rules. While national law requires that affected areas be offset by other sites in a ratio of at least 1:1, the provinces may require larger offset ratios [typically for off-site compensation (Oldenburger 2012)]. Areas that are (newly) afforested for the purpose of wood production can be exempt from compensation upon harvest if harvested within 40 years and with prior consent of the regulator. The Nature Conservation Act of 2017 largely preserves the provisions of

[1] https://www.bij12.nl/onderwerpen/natuur-en-landschap/natuurwetten-en-regelgeving/nederlandse-wetten-en-regelingen/boswet/

Table 1 Key developments of no net loss and offsetting in the Netherlands over the last 10 years

Year	Development	Reference
2007	Publication of 'Spelregels EHS', a policy framework that clarifies which developments are allowed in EHS sites, under what conditions, how impacts should be assessed, principles for offsetting, and designation of offset sites as part of the network. The main principles are outlined in the text (Sect. 2.3)	Ministeries LNV, VROM and provincies (2007)
2007–2010	The National Court of Audit concludes that compliance with mandatory offsets is poor and proper governmental oversight is lacking. The rules and regulations around the topic are complex. Other reports come to similar conclusions	Algemene Rekenkamer (2007)
2009	Establishment of a Taskforce Biodiversity and Natural Capital. The taskforce produced a number of advices, in the context of 'green growth', and also assessed the potential of offsets/habitat banking as an economic instrument to realize green growth (published by De Bie and Van Dessel 2011)	Taskforce Biodiversiteit en Natuurlijke Hulpbronnen (2011) De Bie and Van Dessel (2011)
2010–2013	A National Steering Committee for Biodiversity Offsets studies how biodiversity offsetting rules and regulations in the Netherlands can be designed and applied more efficiently and effectively. The committee asked for a background study to identify current problems and opportunities as well as offset metrics (Broekmeyer et al. 2011). Specifically, the team had the task of determining whether more flexibility in offset requirements (off-site and out-of-kind) could increase uptake and effectiveness. Broekmeyer et al. (2011) concluded that, given the current state of biodiversity, more flexibility would lead to a net loss, and solutions should be sought in clarification (not simplification) of the rules and strengthening of the network, for example through a requirement to have offsets in place prior to the impact	Stuurgroep natuurcompensatie (2013) Broekmeyer et al. (2011)

(continued)

Table 1 (continued)

Year	Development	Reference
2012	Publication of a decree on general rules regarding spatial planning ('Barro'). Chapter 2.10 on the National Nature Network determines that the provinces are obliged to develop rules to ensure no net loss for NNN sites in terms of the habitat quality, area and connectivity. NNL should be ensured by prohibiting activities that significantly affect a site's values, except in cases where: the development is of major public interest; and there are no realistic alternatives; and the negative effects on a site's values, area and connectivity are minimized as far as possible and residual impacts are adequately offset. In the years to follow, provinces have developed such rules (termed 'Uitvoeringsregeling natuurcompensatie'), which differ somewhat from province to province	Link to Barro: https://zoek.officielebekendmakingen.nl/stb-2012-388.html
2012	Preliminary investigations for the development of habitat banking in the Netherlands, requested by the Taskforce on Biodiversity and Natural Capital (established in 2009)	De Bie and Warmerhoven (2012)
2013	'Natuurpact'—a set of agreements between the central government and the provinces regarding nature conservation targets, and how to realize them. These include the larger role for the provinces in governing the NNN network. Moreover, an agreement was made for a programmatic approach to reduce nitrogen levels in Natura 2000 sites (PAS), in order to meet conservation targets and allow economic development. (http://pas.natura2000.nl/pages/home.aspx)	http://www.ipo.nl/files/3313/7949/8317/Hoofdlijnennotitie.pdf
2013, 2014	Investigations for a metric system for offsetting based on credits ('natuurpunten') by the National Fund for Rural Areas (Groenfonds) and the Netherlands Environmental Assessment Agency (PBL). See Sect. 3.3	Groenfonds (2013) Van Gaalen et al. (2014)

(continued)

Table 1 (continued)

Year	Development	Reference
2014	The National Court of Audit concludes that recognition of mandatory offsets has increased amongst developers as well as regulating authorities thanks to changes in policy (e.g. shift from national to provincial responsibilities) and the clearer guidelines for EHS (NNN) areas as laid down in the 'spelregels EHS'. The roles and responsibilities of the various authorities are, however, insufficiently clear, and there is still a lack of monitoring and registration of offsets projects	Algemene Rekenkamer (2014)
2007, 2013, 2017	Regional courts of audit continue to monitor and report on the quality of offsetting in various provinces	Zuidelijke Rekenkamer: (2009a, b, 2013a, b) Randstedelijke Rekenkamer: (2017a, b, c, d)
2012-	Pilots for habitat banks are underway, albeit often with a much broader aim than simply to meet demand for mandatory offsetting. These aims include voluntary offsetting, social corporate responsibility, CO_2 offsets and public engagement. Bugter et al. (2017) summarises three recent initiatives (with a summary in English). See Sect. 3.2	Bugter et al. (2017)
2017	The new Nature Conservation Act ('Wet Natuurbescherming 2017') comes into force, unifying the variety of nature conservation laws into a single act. While the rules and regulations regarding NNL and offsetting are largely unchanged, the roles and responsibilities are more clearly split between the national government (responsible for international obligations (Birds and Habitats Directives), and large water bodies) and the provinces (responsible for national and provincial nature network)	https://www.government.nl/topics/nature-and-biodiversity/new-law-protects-nature-in-the-netherlands
2019 (expected)	It is planned to fully embed the Nature Conservation Act in the Environment and Spatial Planning Act ('Omgevingswet'). The legal framework of NNL and offsetting is expected to stay the same as under the Wet Natuurbescherming 2017	https://www.government.nl/topics/spatial-planning-and-infrastructure/revision-of-environment-planning-laws

Table 2 Summary of conservation regimes in the Netherlands with biodiversity offset requirements (based on Broekmeyer et al. 2012 and Randstedelijke Rekenkamer 2017a)

	Natura 2000	National nature conservation network (NNN), formerly "EHS"	Forests	Protected species
Original legal framework prior to 2017[a]:	Nature Conservancy Act 1998	'Barro',[b] Spelregels EHS (Ministeries LNV, VROM en provincies 2007) and originally Structuurschema Groene Ruimte (Ministerie van LNV 1995)	Forestry Act (1961)	Flora and Fauna Act 1998
Offsetting mandatory in the case of:	Significant negative effects on the site's Favourable Conservation Status	Significant negative effects on key properties of the site (quality, connectivity, size)	>0.01 ha or more than 20 trees from a tree line	Only implicitly required if the conservation status is negatively affected
Spatial offset requirements:	Near the impacted location (to maintain the spatial cohesion of the network)	Preferably near the impacted location	Preferably near the impacted location	None, as long as 'favourable conservation status is maintained'
Temporal offset requirements:	Prior to the impact	Undefined; determined case-by-case	Within three years after clearing	None, as long as 'favourable conservation status is maintained'
Required offset size (multipliers):	Should ensure no net loss	Should be at least equivalent, with a multiplier based on habitat development time	Should be at least of equal size (1:1)	Determined case-by-case
Type of offset:	Same habitat type	Other habitat type possible	Other habitat type possible	Determined case-by-case
In lieu fees (financial compensation) allowed?	No	Yes	No	No

[a]Since January 2017 all regimes fall under the Nature Conservation Act 2017, which will be merged into the upcoming Environment and Planning Act (expected to come into force in 2019). While most regimes are retained under the 2017 Act, the species conservation regime has become less strict for a number of (threatened) species, and is now closer in intent to the European Birds Directive
[b]https://zoek.officielebekendmakingen.nl/stb-2012-388.html

the Forest Act. Mandatory afforestation is no longer required when deforestation has been undertaken as a measure to maintain or increase a particular habitat type under the Natura 2000 regime.[2] The Forest Act primarily aimed at no net loss of forest area

[2]http://www.bomenrecht.nl/wet-natuurbescherming/

in the Netherlands. In 2005 Nabuurs et al. analyzed land cover maps to show that deforestation rates reached 1470 ha/year for the period 1990–2000. A follow-up study by Oldenburger (2012) examined statistical data on de- and afforestation to conclude that approx. 500 ha/year was deforested in the period 2006–2010. Of this figure, approx. 300 ha/year was not offset, leading to a net loss of forest cover of approx. 200 ha/year.

2.2 *Natura 2000*

The legal protection of Natura 2000 sites as laid down in the EU Habitats and Birds Directives was formalised in the Natuurbeschermingswet 1998 (Nature Conservancy Act 1998). The Natura 2000 conservation regime is the strictest conservation regime in the Netherlands. In line with Article 6(4) of the Habitats Directive, activities that negatively affect the integrity of an SAC/SPA site are only granted permission to proceed on the basis of overriding public interest, in the absence of alternatives, and when appropriate compensatory measures are designed and implemented (see also Broekmeyer et al. 2011; Tucker et al. 2014). The regime can be said to be the strictest in regard to the following conditions: (1) it only applies to projects of overriding public interest; (2) offsets have to be designed and implemented before the impact takes place; (3) no financial compensation is allowed; (4) offsets have to be equivalent in habitat type and located near the impact location in order to maintain the spatial cohesion of the network.

The strict regulations of the Natura 2000 regime have often been criticized within the Netherlands. This is probably best exemplified by the letter from the Dutch Prime Minister Balkenende to the President of the European Commission, José Manuel Barroso, in 2009, in which Balkenende pressed for changes to the EU's nature protection policy. According to Balkenende, Natura 2000 'defeated its purpose' and the strict rules threatened to cripple economic activity. Further, he wrote that Natura 2000 faced so much opposition in the Netherlands that support for nature policy was crumbling. This event marked a period of strong opposition from developers and politicians to the limitations imposed by the country's conservation laws. Budgets for nature conservation were severely cut during the financial crisis, and a national steering committee for biodiversity offsets studied how biodiversity offsetting rules and regulations in the Netherlands could be designed and applied more efficiently and effectively (Broekmeyer et al. 2011; Stuurgroep natuurcompensatie 2013). This not only applied to Natura 2000 sites but also to the National Nature Network (NNN, see Sect. 2.3). Despite the hostility, the rules regarding Natura 2000, no net loss and offsetting were not revised. More flexibility was sought through the introduction of a programmatic approach regarding nitrogen (Table 1, 2013 Natuurpact): High nitrogen levels are one of the key pressures on Dutch Natura 2000 sites. Hence, projects often have significant impacts on Natura 2000 sites, due to increases in nitrogen pollution (traffic, agriculture). To improve the condition of sites and simultaneously ease the permitting of such projects, it was agreed to improve the nitrogen pressure on Natura 2000 sites through restoration measures, reduce emissions and ease the permitting process through a programmatic approach. Moreover, the ambitions for habitat recreation/restoration under the NNN regime, were drastically reduced, given severe budget cuts (see Sect. 2.3).

In response to the outcomes of the Fitness Check of the Bird and Habitats Directives (Milieu, IEEP and ICF 2016), the Dutch Ministry of Economic Affairs (responsible for Natura 2000) asked a group of experts to assess the implementation of the Directives in the Netherlands. These findings, reported in Adams et al. (2017), conclude that the implementation of the Directives in the Netherlands is sound and well thought through, with a clear regulatory framework. The notion that the regulatory framework is nevertheless experienced as strict—recently exemplified through the cases of law "Briels" (ECLI:EU:C:2014:330) and "Orleans" (ECLI:EU:C:2016:583)—is primarily related to the poor conservation status of habitat types in the Netherlands, almost all of which have an 'unfavourable-bad' to 'unfavourable-inadequate' conservation status (EEA 2015). Only 4% of habitat types have a favourable conservation status, the lowest percentage of all EU Member States. This means that even relatively small effects of proposed activities are likely to lead to significant impacts, and thus are prohibited. To reduce the friction between conservation and development, the experts therefore concluded that the solution should not be sought in changing the regulations, but primarily in improving the conservation status of biodiversity through adequate conservation and restoration.

2.3 *National Nature Conservation Network Sites (NNN)*

The Dutch national nature conservation network (NNN, formerly known as the EHS) encompasses all of the country's existing nature conservation areas, whether terrestrial areas or water bodies (North Sea, Wadden Sea, rivers and lakes). These include Natura 2000 sites and their associate regime. In addition, the NNN covers many other parks, reserves and corridors of national and provincial relevance. Parts of the network designated as NNN are still in agricultural use (*'unrealized'*); in these areas habitat creation and restoration measures (*'natuurontwikkeling'*) still have to be implemented. As such, there is a distinction between the realized and unrealized parts of the NNN. The Dutch biodiversity strategy of 1990 established the idea of an 'Ecological Main Structure'(EHS)[3] to halt the loss of biodiversity. The first spatial plans were laid down in the Structuurschema Groene Ruimte (SGR) in 1995. Conservation of NNN sites was established by the Nature Conservancy Act of 1998. Since then the rules and regulations regarding NNL and offsetting have been specified and clarified in various documents such as the 'Spelregels EHS' in 2007, and the 'Barro' in 2012 (Table 1). The national government's Nature Conservation Act (2017) currently provides the legal framework for the protection of NNN sites.

[3]https://www.groeneruimte.nl/dossiers/ehs/home.html

In 1990, the network was planned to encompass 453,500 ha of existing (semi) natural habitat, and was supposed to be extended with ca. 275,000 through restoration, totaling ca. 730,000 ha by 2018. Due to the change of government in 2011 and the financial crisis at that time, the restoration ambitions were reduced with 130,000 ha, aiming at a total network size of 600,000 ha only. In 2013 less severe cuts were agreed upon, that is, a network size of 680,000, but with a deadline for realization in 2027 instead of 2018. Between 1990 and 2014 the area of habitat had grown from 453,500 ha to ca. 547,500.[3]

Since 2012 the central government has gradually shifted responsibility for the NNN to the provinces, only retaining responsibility for Natura 2000 sites (Natuurpact 2013). The Nature Conservation Act 2017 formalizes the role of the provinces in the design of the network and enforcement of the conservation regimes. The provinces must derogate the national law into provincial rules in order to enforce the regimes. This has been achieved in different ways: While certain provinces have taken an active role in directing developments and ensuring cohesion of the network, others have pursued a more tentative position (while still enforcing the law), allowing individual municipalities to primarily direct spatial developments (Randstedelijke Rekenkamer 2017a, b, c, d).

The mitigation hierarchy and no net loss are central principles of the NNN set of regulations. In general, developments in conservation sites are not allowed except in cases of major public interest where there are no realistic alternatives and the negative effects on the site's values, area and connectivity are minimized as far as possible and residual impacts are adequately offset ['Spelregels EHS' and 'Barro', Table 1]. Adequate offsetting means that there should be no net loss of the 'key properties and values' of the site in terms of size, quality (habitat type and quality) and connectivity. Specifically, the provisions in the 'Spelregels EHS' were: (1) Offsets should be realised adjacent or close to the impact site; (2) Offsets should be additional, i.e. offsets should not be used to fulfill any other existing restoration obligations; (3) Active restoration measures should be taken to allow equivalent values to develop at the offset site; (4) In order to ensure spatial cohesion and to achieve equivalence in terms of quality, offset ratios larger than 1:1 may be applicable; (5) To offset the loss in quality due to the substitution of a more mature habitat with a newly restored habitat, offset multipliers are applicable. Four categories are used with habitat restoration time and offset multiplier (Table 3); (6) In special circumstances, it is possible to deviate from these categories and associated multipliers; (7) In cases where in-kind offsets are infeasible (for example because of very long habitat restoration times, or because the specific abiotic conditions are lacking outside the network) then out-of-kind offsets are allowed if they are considered of equivalent biodiversity value. (8) When 100% physical offsetting (restoration offsets) is deemed infeasible within the indicated time frame, financial compensation is allowed for the part (and only that part) that cannot be realised. (9) Regarding losses of aquatic habitats, it can be impossible to create new areas as offsets. In such cases, offsets are to be realised through habitat quality improvements and/or improvements in the spatial cohesion of the network.

Table 3 Offset multipliers and additional restoration fees related to habitat restoration time (Broekmeyer et al. 2011, based on *Spelregels EHS*). The assumed restoration time of habitat types are provided as an annex to the *Spelregels EHS*

Habitat restoration time of impacted habitat	Required offset area	Additional fee for restoration costs
<5 years	1 × the area lost (i.e. no additional area required)	–
>5 years but <25 years	1.33 × the area lost	+ costs for restoration management
>25 years but <100 years	1.67 × the area lost	+ costs for restoration management
>100 years	Assessed on a case-by-case basis	Assessed on a case-by-case basis

While these used to be the general principles for mitigating and compensating impacts on NNN sites, they have been the source of some confusion, in part because the NNN comprises a mixture of zones, encompassing for example Natura 2000 sites, other core areas, buffer zones, corridors, and designated areas that are still in agricultural use. Since 2011 more attention has been paid to the rules, regulations and practice around compensation offsetting (Table 1), helping to clarify the roles and responsibilities of the parties and government bodies involved (Algemene Rekenkamer 2014). The National Court of Audit did note, however, that the plurality in zoning systems and regimes was confusing. For example, no up-to-date national map of conservation zones and offset sites is available to determine impacts and size offsets appropriately (Algemene Rekenkamer 2014).

In lieu fees are accepted in cases where no suitable offsets can be realized (e.g. for very large infrastructural projects). Such fees, which are received and managed by the National Fund for Rural Areas, totaled 169 million € in 2016 (Nationaal Groenfonds 2017). With the shift of responsibilities of the NNN from the national government to the provinces, certain provincial authorities are currently also allowing the payment of financial compensation, banked in so-called Provincial Funds. In line with government policy to create larger and better-connected nature conservation areas, these funds are often pooled to help pay for large restoration projects.

To ensure additionality, the 'Barro' (Table 1) dictated that offset sites had to be explicitly located outside the NNN, and subsequently designated as an NNN site to ensure no net loss of network size. The 'Spelregels EHS' also explicitly stated that offsets should not be used to realise previously planned parts of the network. However, in view of the larger role of the provinces and the financial burden they face in meeting nature restoration targets in the unrealized parts of the NNN, certain provinces specifically demand that restoration offsets be allocated at unrealized sites of the NNN (Fig. 1, element B). This has the benefit of speeding up the creation of these network areas as well as generating larger and better-connected sites locally. However, this should still be accompanied by an extension of the total size of the NNN (Fig. 1, element C). Failure to do so will result in a reduction in the size of the

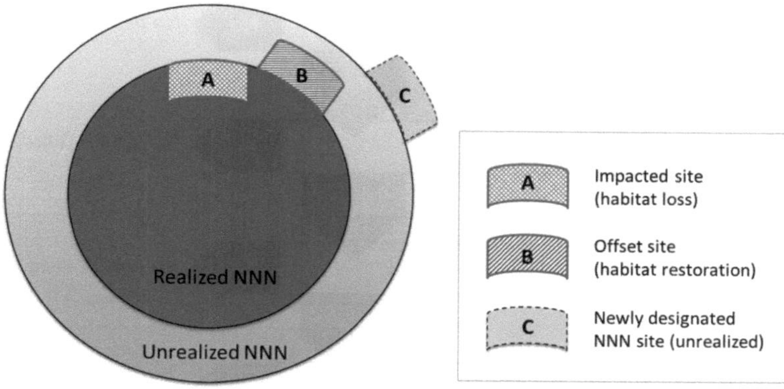

Fig. 1 Graphical representation of ensuring no net loss, when an impact (**a**) on the conservation network (NNN) is offset by restoring a site at the unrealized part of the network (**b**). To ensure no net loss in terms of network size, a new site will have to be added to the unrealized part of the network (**c**). The unrealized NNN is planned to be realized (through habitat creation and restoration) by 2027. *Modified from Randstedelijke Rekenkamer* (2017a, b)

NNN (and thus a net loss) as the impacted site is taken out of the NNN. A regional Court of Audit highlighted this aspect when evaluating the practices of the province of North Holland (Randstedelijke Rekenkamer 2017b).

2.4 Species

Species have been protected in the Netherlands since 1998, initially under the Flora and Fauna Act of that year, later superseded by the Nature Conservation Act of 2017. While the Flora and Fauna Act did not make explicit mention of mitigation and compensation, these could be required implicitly, as exemptions to the law were only permitted on the basis that species maintained their favourable conservation status despite the impact (Broekmeyer et al. 2011). In practice, such conditions under which permits for developments with negative impacts on species were granted were termed 'compensatory measures'.

The new Conservation Act has three categories of protected species: (1) all native European bird species; (2) all other protected European species (i.e. species of the Habitats Directive Annex 4a; the Bern Convention Annex 2 and the Bonn Convention Annex 1); (3) other 'nationally' protected species.

The Ministry of Economic Affairs (2016) has described the impact of the new law with respect to species conservation. In short, for any proposed plan, the steps outlined in Fig. 2 should be followed to ensure compliance with the Nature Conservation Act 2017 regarding protected species. The most important changes to the

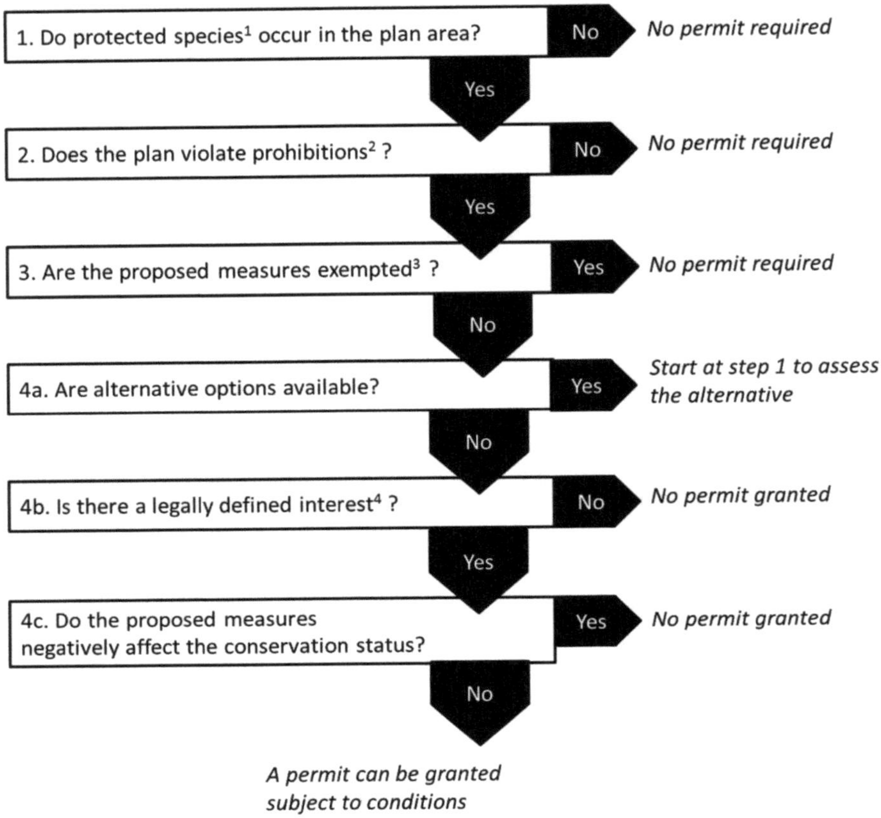

Fig. 2 Sequence of questions to determine whether a plan requires a permit in the context of the species conservation regime [modified from Ministry of Economic Affairs (2016)]. Notes: (1) there are three categories of protected species; (2) each category of protected species has different prohibitions; (3) different exemptions exist depending on the proposed action and species in question; (4) legal interests vary per category of protected species

provisions of the Flora and Fauna Act (1998) (Ministry of Economic Affairs 2016) are as follows:

- The province now has the sole legal authority to grant exemptions and permits.
- The lists of protected species have been revised: Over 100 species have been removed from the lists, while around 80 other species now enjoy protection. In total about 970 species are protected, including over 700 bird species.
- The prohibitions are more closely aligned with the Birds and Habitats Directives. Moreover, certain accidental impacts are no longer considered violations.
- There is a provision to exempt impacts in the context of a programmatic approach (i.e. allowing gains and losses to be balanced at the level of a programme).
- The exemption criteria are more closely aligned with the Birds Directive, for which fewer cases of disturbance to birds require permits.

Although the Ministry of Economic Affairs (2016) states that relatively few changes apply to species protected by the Birds and Habitats Directives (Groups 1 and 2 in the Nature Conservation Act), there is one important revision: Whereas any disturbance of protected species was prohibited under the Flora and Fauna Act, the new law does not consider *accidental* disturbances to be legal violations. Furthermore, for species of national conservation interest (Group 3), disturbances are no longer considered to be violations (such disturbances were prohibited under the Flora and Fauna Act). Disturbances of individual birds is no longer prohibited if the disturbance has no significant influence on the conservation status of the species. However, cumulative effects must be considered and the destruction of nests during the breeding season remains prohibited (Ministry of Economic Affairs 2016).

Reflecting the Flora and Fauna Act, the species conservation regimes in the Nature Conservation Act do not specifically distinguish between mitigation and compensation measures. Nevertheless, the design of the law promotes avoidance. In cases where permits are granted, conditions typically apply that mitigate and/or offset the impacts on the species' conservation status.

3 Practice in the Netherlands

While policies regarding the mitigation hierarchy and offsetting have been in place for decades, these have been poorly realized in terms of application, enforcement and monitoring (Algemene Rekenkamer 2007; Zuidelijke Rekenkamer 2009a, b; Broekmeyer et al. 2011). According to the Steering Committee on Biodiversity Offsets (2013), there are four main reasons for this: (1) the complexity of the legislation; (2) the diversity and multitude of conservation regimes and offset regulations; (3) differences between the provinces; and (4) limited knowledge of offsetting requirements and regulations. In this chapter we briefly outline a number of key initiatives to improve conservation practice in addition to the changes/clarifications in rules and regulations already outlined in Sect. 2.

3.1 Towards More Transparency in the Application and Monitoring of Offset Projects

In the Netherlands, offsetting activities are linked to permits to ensure compliance; that is, a planning permit is only granted after the developer has provided authorities with a plan detailing how the offset will be implemented. Compliance with the compensation requirements is enforced through fines if activities are not undertaken in accordance with the agreed offset provision plan. Each province has to monitor the state of compliance. While adequate compensation is strictly enforced under the Birds Directive and Habitats Directive, the National Nature Network compensation scheme has so far been weakly enforced (Algemene Rekenkamer 2007; Zuidelijke Rekenkamer 2009a,b; de Bie, pers. comm. in Tucker et al. 2014). The National Court of Audit recently concluded that offsetting practices have improved since 2007 thanks to clarifications of roles and responsibilities as well as reduced complexity, yet also found that provinces still have insufficient insight or overview of the offsetting requirements when permits are granted. The lack of general guidelines for registration leads to gross variations in the process, and hinders the comparison of data across provinces (Algemene Rekenkamer 2014). Ostensibly, information on all offsets in the Netherlands is available online at http://www.ruimtelijkeplannen.nl, with links to all individual spatial plans and permits. Extracting that information, however, means scrutinizing the documentation on a plan-by-plan basis. This is hindered by the fact that only a fraction of all spatial plans and permits available through the webportal involve offsetting, and the inability to generate a project list. Although provinces are required to compile overviews of offsets projects on an annual basis and to monitor offsets, these overviews are not generally available to the public (Bull et al. 2018). Improvements are, however, being made in this regard since the publishing of the National Court of Audit report.

3.2 Habitat Banking Pilots

For over 10 years various actors in the Netherlands have expressed interest in habitat banking, such as the National Fund for Rural Areas, the Ministry of Environmental Affairs, the Platform Business, Biodiversity and the Economy (Platform BEE) as well as several provinces. There is a diverse range of motivations. First, the government's aim is to reduce friction between conservation and development, and to achieve a better balance between the demand and supply of offsets (e.g. Stuurgroep natuurcompensatie (2013); see also www.bosennatuurcompensatie.nl/ for an initiative to fulfill a brokerage role between demand and supply). Second, while compensation funds as held by the National Fund for Rural Areas (Nationaal Groenfonds) have to be spent on offset projects, it is a challenge to find suitable locations (G. van der Vliet, Nationaal Groenfonds, pers. comm 2008). Third, developers wish to ease the approval process and avoid the need for offsets through the facilitation of restoration measures up-front (Adams et al. 2017) as well as to foster Social Corporate Responsibility (De Bie and Warmerhoven 2012). De Bie and Warmerhoven (2012) have assessed the potentials for a system of habitat banking in the Netherlands, while Bugter et al. (2017) have described and evaluated three recent habitat banking systems, including one developed by De Gemeynt in line with the proposals of De Bie and Warmerhoven (2012). Bastmeijer and Van Kreveld (2017) have described a compensation pool initiative in the province of Gelderland as an example of a more proactive nature conservation policy, in contrast to the somewhat reactive and defensive policy practiced for some time in the Netherlands. Bugter et al. (2017) conclude that:

> In principle, private parties appear to be interested in investing in the quality of the environment, including nature and landscape. The motive for possible investments is the general social desire to manage the environment in a better, more sustainable way. For companies it is therefore both useful and necessary to visibly invest in sustainability. In this respect, private parties have a wide investment interest, which is much more motivated by concern for the quality of the environment, with nature and landscape as a part of it, than by ecology or biodiversity. Some of the things mentioned as worth investing in are: recreation (facilities), improving the quality of life, agricultural areas and planting for CO_2 compensation. Important factors determining the willingness to invest seem to be the effort needed and the returns on the investment. The latter may take various forms. Most often mentioned were: options for spatial development/expansion space, preferential treatment for tenders, certification and PR purposes. This may also mean that companies want to make the investment in their own region and that they want some measure of control over how their money is invested.

Further, Bugter et al. (2017) identify a number of requirements for successful habitat banking, including: (1) an appropriate, transparent and above all flexible system of rewards and returns; (2) regulations and a regulatory organisation which must have a much broader base than a traditional offset bank, but which could incorporate such an offset bank if required. These requirements reflect the findings of Van Teeffelen et al. (2014), who synthesized ecological, economic and institutional requirements for conservation banking in intensively used landscapes such as the Netherlands.

3.3 Development of a Point-System-Metric

In 2013 the Dutch Fund for Rural Areas proposed a point-system-metric ('natuurpunten') to balance losses and gains in habitat quality and quantity, to promote the identification of appropriate offsets, and to ease the approval process (Nationaal Groenfonds 2013). While the proposed method is summarized in the following, any assessment of its efficiency is beyond the scope of this chapter. To our knowledge, the metric has not yet been adopted in practice. Parts of the methodology have been described in or proposed by Broekmeyer et al. (2011) and De Bie and Warmerhoven (2012). The Netherlands Environmental Assessment Agency has studied the application of such an approach to aquatic systems (Van

Gaalen et al. 2014), and Vergouwen et al. (2015) has built on these methods with the aim to better align infrastructural projects with nature conservation policy.

Following the methodology presented in the report by the Nationaal Groenfonds (2013), ecological value is determined by three factors using the equation:

$$\text{Ecological value} = \text{Habitat Area} \times \text{Habitat Quality} \times \text{weighting factor}$$

The three factors are defined as follows:

Habitat Area Every hectare of habitat lost or gained is assigned one point.

Habitat Quality Referred to as ecological quality, this is a measure of the actual (impacted area) or potential (offset area) 'completeness' of the ecosystem in terms of species composition, which is supposedly assessed as the difference between the potential and actual species composition of a habitat site.

- **Actual species composition** is based on data from the national database on flora and fauna or the national vegetation database, complemented by field surveys where needed.
- **Potential species composition** is determined using ecological suitability maps. According to the report, these are available for all habitat types (1×1 km) and for 160 of 280 species protected through the Flora and Fauna Act. Subsequently, the Mean Species Abundance (MSA) index can be applied to calculate the ecological quality. The list of key species per habitat type provides a reference for optimal conditions (available for all habitat types eligible for subsidies under the 'SNL regeling'). For other habitats, the ecological quality is derived from the vegetation type.

Weighting Factor This is determined by considering the legal status of the site/habitat/species concerned, the habitat restoration time and degree of habitat connectivity.

- Legal status/policy status habitats: higher status = more points. Status is determined by so-called 'ITZ criteria', where I = international importance, T = trend and Z = rarity ('zeldzaamheid'). These criteria are similar to the criteria adopted by the EU Habitats Directive.
- Legal status/policy status species: habitats of protected species are assigned more points on the condition that (vital) populations are present, i.e. not just individuals.
- Restoration time: longer time = more points.
- Connectivity: the degree to which the location contributes to the connectivity of the network. The larger the contribution, the more points.

Although the report includes examples, details are not given on how to determine the actual points allocated to the different aspects. The approach is further elaborated in Vergouwen et al. (2015, available in English).

4 Conclusion

In the Netherlands the legal provisions necessary to conserve habitats and species through the mitigation hierarchy and offsetting are in place and are being applied. While many changes to the rules, roles and responsibilities around offsets have complicated their application over the last decade, recent audits have indicated some improvement in this regard. Today, initiatives and discussions focus on: (1) improving the transparency of projects and their offset requirements in order to facilitate the monitoring, enforcement and application of the legal framework; and (2) ways of applying the principles of habitat banking and associated metrics to create a more robust network that could reduce the significance of impacts, thereby increasing flexibility in the conservation regimes and easing the approval process.

References

Adams AS, Backes CW, Drahmann A (2017) Een betere implementatie van de VHR in Nederland—Bevindingen van experts. https://www.rijksoverheid.nl/documenten/rapporten/2017/03/01/een-betere-implementatie-van-de-vhr-in-nederland

Algemene Rekenkamer (2007) Bescherming Natuurgebieden. Den Haag, the Netherlands. Available https://zoek.officielebekendmakingen.nl/kst-31074-2.pdf

Algemene Rekenkamer (2014) Compensatie van schade aan natuurgebieden: vervolgonderzoek naar de bescherming van natuur gebieden. Den Haag, Netherlands. Available https://www.rekenkamer.nl/publicaties/rapporten/2014/04/16/compensatie-van-schade-aan-natuurgebieden

Bastmeijer K, Van Kreveld A (2017) Meer Natuur Minder Knel. Met de Wet natuurbescherming naar actief natuurherstel. Bureau Ulucus & Legal Advice for Nature. https://www.landschappen.nl/Uploaded_files/Zelf/meernatuurminderknel.dedcdf.pdf

Broekmeyer MEA, Bugter RJF, van Teeffelen AJA (2011) Natuurcompensatie: kansen, knelpunten en rekenregels. Alterra. http://edepot.wur.nl/188326

Broekmeyer MEA, Bugter RJF, Van Teeffelen AJA (2012) Natuurcompensatie: slim beleid begint op tijd. Landschap 29(2):57–67

Bull JW, Brauneder K, Darbi M, Van Teeffelen AJA, Quetier F, Brooks S, Dunnett S, Strange N (2018) Data transparency regarding the implementation of European 'no net loss' biodiversity policies. Biol Conserv 218:64–72. https://doi.org/10.1016/j.biocon.2017.12.002

Bugter R, Vader J, van den Hoven M (2017) Investeren in natuur: Naar pilots voor een bredere vorm van habitatbanking. Wageningen Environmental Research, Wageningen. https://doi.org/10.18174/413468

De Bie S, van Dessel B (2011) Compensation for biodiversity loss. Advice to the Netherlands' Taskforce on Biodiversity and Natural Resource. De Gemeynt. http://www.gemeynt.nl/files/Pb2011-002.pdf

De Bie S, Warmerhoven H (2012) Voorstudie Habitatbanking – voorstellen voor de systematiek, organisatie en uitvoering van habitatbanking in Nederland. De Gemeynt U.A., Klarenbeek. http://www.platformbee.nl/assets/Pb2012-001_Voorstudie_Habitatbanking.pdf

EEA (2015) State of nature in the EU. Results from reporting under the nature directives 2007–2012. European Environment Agency. Technical report No 2/2015

Ministry of Economic Affairs (2016) Soortenbescherming bij ruimtelijke ingrepen. Ministry of Economic Affairs, Den Haag. https://www.synbiosys.alterra.nl/natura2000/documenten/effectenindicator/Soortenbescherming_bij_ruimtelijke_ingrepen_1.4_01022017.pdf

Ministeries LNV, VROM, provincies (2007) Spelregels EHS – Beleidskader voor compensatiebeginsel, EHS – saldobenadering en herbegrenzen EHS. Den Haag. https://www.rijksoverheid.nl/documenten/rapporten/2007/08/20/spelregels-ehs

Milieu, IEEP and ICF (2016) Evaluation study to support the fitness check of the birds and habitats directives. http://ec.europa.eu/environment/nature/legislation/fitness_check/docs/study_evaluation_support_fitness_check_nature_directives.pdf

Nabuurs GJ, Kuikman PJ, Kramer H (2005) Nederlandse ontbossing bedraagt 1470 ha per jaar. Vakblad Natuur Bos Landschap 2(7):27

Nationaal Groenfonds (2013) Puntensysteem natuurcompensatie en – saldering. http://natuurlijkkapitaal.com/wp-content/uploads/2016/05/Puntensysteem-natuurcompensatie-en-saldering-Groenfonds.pdf

Nationaal Groenfonds (2017) Jaarverslag 2016. Hoevelaken. https://www.nationaalgroenfonds.nl/over-groenfonds/jaarverslagen

Oldenburger J (2012) Bosomvorming zonder compensatie. Stichting Probos, Wageningen. http://www.probos.nl/rapporten

Randstedelijke Rekenkamer (2017a) Natuurlijk bevoegd: onderzoek naar de invulling van de provinciale rol bij natuurcompensatie. Provincie Flevoland. http://www.randstedelijke-rekenkamer.nl/onderzoek/natuurcompensatie/

Randstedelijke Rekenkamer (2017b) Natuurlijk bevoegd: onderzoek naar de invulling van de provinciale rol bij natuurcompensatie. Provincie Noord Holland. http://www.randstedelijke-rekenkamer.nl/onderzoek/natuurcompensatie/

Randstedelijke Rekenkamer (2017c) Natuurlijk bevoegd: onderzoek naar de invulling van de provinciale rol bij natuurcompensatie. Provincie Zuid Holland. http://www.randstedelijke-rekenkamer.nl/onderzoek/natuurcompensatie/

Randstedelijke Rekenkamer (2017d) Natuurlijk bevoegd: onderzoek naar de invulling van de provinciale rol bij natuurcompensatie. Provincie Utrecht. http://www.randstedelijke-rekenkamer.nl/onderzoek/natuurcompensatie/

Stuurgroep natuurcompensatie (2013) Eindrapport Stuurgroep natuurcompensatie, vol 11. Den Haag. https://zoek.officielebekendmakingen.nl/blg-253318.pdf

Taskforce Biodiversiteit en Natuurlijke Hulpbronnen (2011) Groene groei – investeren in biodiversiteit en natuurlijke hulpbronnen. www.natura2000.nl/files/eindrapport_taskforce_biodiversiteit.pdf

Tucker G, Allen B, Conway M, Dickie I, Hart K, Rayment M, Schulp C, van Teeffelen A (2014) Policy options for an EU no net loss initiative. Institute for European Environmental Policy, London. http://ec.europa.eu/environment/nature/biodiversity/nnl/index_en.htm

van Gaalen F, van Hinsberg A, Franken R, Vonk M, van Puijenbroek P, Wortelboer R (2014) Natuurpunten: kwantificering van effecten op natuurlijke ecosystemen en biodiversiteit in het Deltaprogramma. PBL, Den Haag. http://www.pbl.nl/sites/default/files/cms/publicaties/PBL_2014_Natuurpunten_Achtergrondstudie_1263.pdf

van Ministerie LNV (1995) Structuurschema Groene Ruimte – Het landelijk gebied de moeite waard, Deel 4: Planologische kernbeslissing. Ministeries van LNV en VROM, Den Haag

van Teeffelen AJA, Opdam P, Wätzold F, Hartig F, Johst K, Drechsler M, Vos CC, Wissel S, Quétier F (2014) Ecological and economic conditions and associated institutional challenges for conservation banking in dynamic landscapes. Landsc Urban Plan 130:64–72. https://doi.org/10.1016/j.landurbplan.2014.06.004

Vergouwen S, Moinier S, Beumer V (2015) How do infrastructural designs meet up to nature policy requirements? A method to evaluate infrastructural (eco-engineering) designs in the light of nature policies. Deltares. http://publications.deltares.nl/1209423_007.pdf

Zuidelijke Rekenkamer (2009a) Kwaliteit Natuurcompensatie provincie Noord Brabant. Deel II Rapport van bevindingen. http://www.zuidelijkerekenkamer.nl/89/archief/noord-brabant.html

Zuidelijke Rekenkamer (2009b) Kwaliteit Natuurcompensatie provincie Limburg. Deel II Rapport van bevindingen. http://www.zuidelijkerekenkamer.nl/91/archief/limburg.html

Zuidelijke Rekenkamer (2013a) Kwaliteit Natuurcompensatie provincie Noord Brabant (Vervolgonderzoek). Deel II Rapport van bevindingen. http://www.zuidelijkerekenkamer.nl/89/archief/noord-brabant.html

Zuidelijke Rekenkamer (2013b) Kwaliteit Natuurcompensatie provincie Limburg (Vervolgonderzoek). Deel II Rapport van bevindingen. http://www.zuidelijkerekenkamer.nl/91/archief/limburg.html

United Kingdom

Jonathan Baker, Liza Papadopoulou, and William Sheate

Before the mid-2000s biodiversity offsetting was not part of the environmental policy discourse in the United Kingdom (UK). Since that time, of the UK's four national administrations, England has been the most pro-active with regard to biodiversity offsetting.[1] As a result this chapter focuses on the policy and practice of England with a short section summarising relevant developments in Wales, Scotland and Northern Ireland.

Within this chapter, offsite compensation/biodiversity offsetting required as part of the EU Birds and Habitats Directives (transposed to Regulations in the UK) is excluded. This is because practice within England is broadly consistent with the rest of Europe.

The requirement to account for, and where possible reduce, the impact of development[2] on biodiversity is a long-standing part of spatial planning and development control across the UK. Over this time mitigation and compensation has tended to be ad hoc with few standards in place or tools to support the quantification

[1]Environmental policy is a devolved matter so the Devolved Administrations of Wales, Scotland and Northern Ireland are responsible for developing their own policy; within a framework set by the United Kingdom Government.

[2]Note in this context development refers to the scope of the Town and Country Planning Act 1990 (as amended). In summary this includes all changes in land use associated with the creation or alternation of industrial, commercial, residential and infrastructure. Changes in agricultural land management are dealt with separately. Nationally significant infrastructure is also excluded.

J. Baker (✉)
London, UK
e-mail: jonathan-baker@hotmail.co.uk

L. Papadopoulou
Fairfax, VA, USA
e-mail: liza.p.323@gmail.com

W. Sheate
Collingwood Environmental Planning, London, UK
e-mail: w.sheate@cep.co.uk

W. Wende et al. (eds.), *Biodiversity Offsets*,
https://doi.org/10.1007/978-3-319-72581-9_12

of impact. Within England biodiversity offsetting differs from 'business as usual' in that it requires the quantification of biodiversity impacts and the commitment to demonstrate net gain or at least no net loss of biodiversity. These requirements are generally missing from 'normal' compensation practice within England.

For instance, the evidence suggests that although 40% of all local planning authorities have experience with some form of offsite compensation (Newey 2012) as few as 1 in 1000 planning applications are required to compensate for any biodiversity lost (Baker et al. 2014a). In those instances where compensation occurs, on average the loss of 1 ha of developed land is compensated for by 0.58 ha of offset (Newey 2012).

Since the mid-2000s the Government Department responsible for the natural environment, the Department for the Environment Food and Rural Affairs (DEFRA), has been exploring whether and how biodiversity offsetting might contribute to England's biodiversity targets. This has included developing technical materials, including a habitat based metric, which were piloted in several parts of the country. Current policy in England is referred to as 'permissive' meaning that the central Government has provided some structures and resources which are available should local authorities and developers choose to use biodiversity offsetting.

1 Policy in England

This section introduces the policy relevant to biodiversity offsetting, in so far as it exists, and describes recent and ongoing changes in approaches to biodiversity mitigation and compensation.

1.1 Legal and Planning Background in England

In the last decade, policy developments in the UK, including the Countryside and Rights of Way Act (2000), the Natural Environment and Rural Communities (NERC) Act (2006), the Town & Country Planning Act (1990, as amended) and the National Planning Policy Framework (NPPF) (DCLG 2012), have allowed planning authorities to seek compensation for impacts on a broader spectrum of biodiversity than was previously the case (EMTF 2013).

Specifically, the NERC Act explicitly places a "*duty to conserve biodiversity*" on every public authority, defining conservation as "*restoring or enhancing a population or habitat*" (Section 40). Reference is made to local authorities as statutory undertakers with regard to their role in planning applications: "*Every public authority must, in exercising its functions, have regard, so far as is consistent with the proper exercise of those functions, to the purpose of conserving biodiversity*". Local Planning Authorities (LPA) which are the public body tasked with shaping planning policy and approving (or not) development—have a statutory duty to consider biodiversity protection when granting planning permission. This duty is open to interpretation, in particular with respect to the requirements for enhancement and compensation for any residual adverse impacts of any given development. Furthermore, the Act requires the Secretary of State to take further steps to conserve and encourage others to take similar actions (see Section 41 of the Natural Environment and Rural Communities Act 2006).

England has a relatively centralized planning system whereby a strategic framework—expressed through guidance and supported by legislation—is created at the national level and then implemented by LPAs, of which there are more than 300. In this context, LPAs have two main functions: creating a plan or set of plans that set out where and in what circumstances development will be permitted; and, responding to developers' applications for planning permission. This local implementation of national guidance is overseen by 'planning inspectors' who review the local plans and decisions regarding planning permission to ensure consistency with national guidance. If a developer is refused permission they have a right to appeal, at which time a planning inspector will review the application. In most instances, the final decision regarding planning permission is made by locally elected councillors. But their decisions must be legally sound—as determined by the planning inspectors and/or the Courts.

The resulting 'plan based system' is intended to be a series of nested plans and frameworks that proceed from the national level strategies to individual developments. This system sets out how biodiversity, among many other factors, will be considered and what mitigation and compensation activities are considered appropriate and necessary for planning permission to be granted.

Consistent with much of Europe, environmental assessment generally and Environmental Impact Assessment (EIA) specifically is an established part of how biodiversity is considered in the planning system. Although it is estimated that less than 0.001% of all planning applications in England require it (IEMA 2011; ONS 2012).[3] EIA has a major role to play in how biodiversity is considered in the planning system as, by definition, those 0.001% of applications have a likely significant impact. Secondly, environmental assessment introduced the concept of the "mitigation hierarchy" when considering the impact of developments. This hierarchy has become part of national planning guidance (Baker et al. 2014b).

1.1.1 Mitigation Hierarchy in England

The mitigation hierarchy is intended to ensure that development activities do not have an unnecessarily negative impact on biodiversity and natural systems (and potentially non-environmental although this chapter only considers environmental aspects). Within England the mitigation hierarchy is embedded in EIA and national planning policy (Department for Communities and Local Government 2012; Parliamentary Office of Science and Technology 2013). The interpretation of the mitigation hierarchy within English legislation, planning applications and planning

[3]Page 19 estimates of ONS (2012) states there are 500 EIAs per year in the UK.

guidance determines the extent to which biodiversity offsetting is possible. Currently the basic elements are in place with the terms compensation and offset being part of relevant legislation and guidance. For instance, the mitigation hierarchy is recognised in the EIA Directive (consolidated Directive following 2014/52/EU amendment) Article 5 (1) (c):

> . . . a description of the features of the project and/or measures envisaged in order to avoid, prevent or reduce and, if possible, offset likely significant adverse effects on the environment;

In this case "remedy" includes compensation, but could also include restoration/rehabilitation.

The phrase has also been transposed into the Town and Country Planning EIA Regulations (2011) SI 1824 Schedule 4, Part 1, Para. 5 as:-.

> A description of the measures envisaged to prevent, reduce and where possible **offset** any significant adverse effects on the environment (emphasis added)

These two legislative definitions have informed England's statutory planning guidance—the National Planning Policy Framework (NPPF); paragraph 118 of which includes the following direction:

> if significant harm resulting from a development cannot be avoided (through locating on an alternative site with less harmful impacts), adequately mitigated, or, as a last resort, compensated for, then planning permission should be refused

The NPPF was introduced in 2010 and although the level of detail within the NPPF is less than in the preceding guidance document (Planning Policy Statement 9), the NPPF maintains the principles that were in place (Newey 2012). Regarding compensation there is a change of emphasis; the NPPF states that compensation "*is a last resort*" something which is missing from PPS9. The idea that compensation and any biodiversity offsetting should be a last resort is an important part of the debate about biodiversity offsetting in England as many conservationists had, and continue to have, concerns about it being a 'licence to trash' (Sullivan and Hannis 2015). The inclusion of last resort in the NPPF seeks to address this somewhat.

It should be noted that the normative definition of the mitigation hierarchy, where on-site compensation is prioritised above off-site (as observed in Baker et al. 2014b), is not formalised in policy or practice guidance. As a result, some have looked to push all compensation off-site via biodiversity offsets to increase the 'net-developable area' of a development. In the 2013 Green Paper (DEFRA 2013a) this increase in net-developable area was included as a potential benefit of biodiversity offsetting. In their consultation response, many stakeholders were highly critical of this (DEFRA 2013b). Others have reflected that onsite compensation is often poorly managed and ineffective management is rarely enforced and that therefore at least having the option of offsite compensation offers the potential for improved biodiversity outcomes (Drayson and Thompson 2013). In the absence of any explicit change in policy; the current assumption, that all options for onsite compensation will be utilised before offsite is considered, remains.

Paragraph 9 of the NPPF states that achieving sustainable development includes "*moving from a net loss of bio-diversity to achieving net gains for nature*;" and paragraph 118 goes on to state that "*the planning system should contribute to ... minimising impacts on biodiversity and providing net gains in biodiversity where possible*." The definition of the mitigation hierarchy in the NPPF and the requirement for net gains form the basis for compensation within England.

Consistent with the streamlined ethos of the NPPF there is no additional information about how 'significant harm' might be determined; this is to be arrived at by the local authorities overseen by the Planning Inspectorate. In practice, significant harm is interpreted in quite a narrow way; with sites protected under national and European biodiversity designations used as a proxy for significance. The online Planning Policy Guidance supports the NPPF and does include the following additional guidance with regard to how mitigation or compensation measures can be ensured: "*the applicant might offer a scheme tailored to the specific context, or consider the potential for biodiversity offsetting with the local planning authority*" (Department for Communities and Local Government 2015). This is consistent with the voluntary approach currently in place within England.

It is within the space created by national guidance which is equivocally supportive and its interpretation by local authorities, that biodiversity offsetting has, to date, existed within England.

English planning recognises the biodiversity value of protected areas and creates a higher threshold for development (as set out in paragraph 118 of the NPPF). This includes international designations such as Special Protection Areas/Special Areas of Conservation as well as national designations such as Areas of Outstanding Natural Beauty (AONB), National Parks, ancient woodlands and Sites of Special Scientific Interest (SSSI). Other national designations such as Local Wildlife Sites and National Nature Reserves are not explicitly prioritised in national guidance.

Current policy is that any compensation for loss of international designations should be managed through the mechanisms set out in the Habitat and Birds Directives rather than through biodiversity offsetting.

The following mechanisms could be used to manage any offsets.

- Section 106 agreements
  - Section 106 of the Town and Country Planning Act 1990 allows a local Planning Authority to enter into a legally binding agreement or planning obligation with a landowner in response to the granting of planning permission. The majority of s106 agreements are not related to conservation or biodiversity.
- Community Infrastructure Levy (CIL)
  - The emerging CIL does have the potential to allow for offsetting if local authorities incorporate biodiversity within the definition of 'infrastructure' (which is defined very broadly and can include green infrastructure) and include it in their CIL Methodology Documentation. Funds that are paid into CIL are not hypothecated which makes it less suitable for biodiversity

offsetting as there is no process to ensure that the compensation monies are spent on a suitable offset.

- Planning Conditions
  - Could be used to require certain measures to be in place prior to development commencing or being occupied. So, for instance a planning condition might require that the development can commence when delivery measures and funding related to delivering offsets are in place and the LPA has confirmed they are sufficient. Planning conditions cannot be used to implement biodiversity offsetting per se, rather they can be used to ensure appropriate system are in place prior to work commencing.
- Compensation via the Nature Directives
  - Due to the high level of protection afforded to Natura 2000 sites, compensation has been used on limited and exceptional occasions to compensate for impacts to sites protected under the Habitats and Birds Directives. Appropriate Assessment (AA) is the analytical mechanism required to determine potential impacts on Natura 2000 sites and the need for compensation. Residual impacts that require compensation are only permissible where there is no alternative and there are imperative reasons of overriding public interest. Biodiversity offsetting is not intended—as designed by DEFRA and the implementation of the metric—to apply to impacts on Natura 2000 sites. It would also be inappropriate for offsetting for impacts elsewhere to occur on Natura 2000 sites as they would be unlikely to provide additional nature conservation benefits beyond those already required under the Directives. In practice where Natura 2000 sites are adversely affected compensation is administered in line with the Birds and Habitat Directives, rather than through the resources developed for biodiversity offsetting.
  - The Government's environmental regulator, Natural England, has brought through some changes to the compensation of European Protected Species[4]; notably the Great Crested Newt which is rare across Europe but in abundance across much of England. The amended policies mean it is easier to damage populations of Great Crested Newts as long as activities were implemented to ensure the local population was not adversely affected; part of this policy includes offsite compensation.
- Environmental Liability Directive (ELD)
  - Though primarily interested in unintentional environment impacts (as opposed to EIA/SEA and AA which are more ex ante assessment tools) the ELD allows for the *compensatory remediation measures* (Annex II) to restore natural resources to their baseline conditions.

[4]Which are protected via Article 12, Annex IV of the Birds and Habitats Directives.

Experience to date suggests that funding secured through Section 106 combined with planning conditions (which stipulates a set of conditions under which planning permission is provided) is the most appropriate mechanism for delivering offsets.

1.2 *Methodological Development in England*

This section describes the policy and methodological development of biodiversity offsetting in the UK to date.

1.2.1 Policy Development

In 2007 the UK Government's Department for the Environment Food and Rural Affairs (DEFRA) identified a need to explore new policy options to halt the continuing loss of non-designated biodiversity; this included potential for mechanisms "*such as biodiversity offsets*" (DEFRA 2007). In 2009 DEFRA commissioned a scoping study which considered the appropriateness, desirability and feasibility of biodiversity offsetting within England. The study highlighted the possible benefits of internalising the costs of development as a means of generating a stream of investment towards the natural environment. The study concluded that although English planning and nature conservation policy has some of the features necessary for biodiversity offsetting, additional mechanisms, tools and policies were needed if it was to become established and effective (Treweek et al. 2009).

In response to the concerns raised from environmental and developer stakeholders, DEFRA has looked to take a considered, evidence-based approach which has seen the development of a significant evidence base for evaluating the efficacy of biodiversity offsetting (DEFRA 2011a).

Biodiversity offsetting was part of the Conservative Party's 2010 Green Paper Open Source Planning—described there as conservation credits (Conservative Party 2010). The Conservative Party set out a vision of biodiversity offsetting which "*reflect(ed) the importance of diverse ecosystems to our environment*". The Conservative Party was elected to Government as part of a coalition with the Liberal Democrats and in 2015 was elected as the solitary party of Government. Since 2010 a Conservative politician has been Secretary of State for the Environment, Food and Rural Affairs and in charge of DEFRA. Opposition parties have not clearly stated whether or not they are supportive of biodiversity offsetting. The exception is the Green Party whose 2015 Manifesto included a promise to stop any biodiversity offsetting (Green Party 2015).

Since 2010 there have been a number of developments related to biodiversity offsetting, these are summarised here:

Development	Date(s)	Description
Making Space for Nature: A review of England's Wildlife Sites	September 2010	In 2010, Making Space for Nature: A review of England's Wildlife Sites and Ecological Network, commonly referred to as the Lawton Review (Lawton et al. 2010) recognised that "*protecting and managing components of England's ecological network is essential, but will not be sufficient to make the network resilient and coherent*". The report was highly influential—for instance being heavily cited in the Government's subsequent Natural Environment White Paper (NEWP)—and identified biodiversity offsets as a potential new funding mechanism for habitat creation and a conservation tool that can assist in materialising the aspirations of the Lawton review for the creation of a nation-wide ecological network
Options Stage Impact Assessment published	June 2011	DEFRA and Natural England developed an impact assessment to set out the options for biodiversity offsetting policy (HM Government 2011a). The three options considered were: 1. Roll out a national scheme mandating use of offsetting for developments impacting biodiversity 2. Roll out a national voluntary scheme encouraging the use of offsetting for developments impacting biodiversity 3. Test out option 2, working with Local Planning Authorities and developers to refine and test the proposals for a voluntary approach The Impact Assessment suggested that the benefits of taking a consistent approach to biodiversity offsetting to implement the requirements of planning policy outweigh the costs. However, it concluded that insufficient information was available to make the case for implementing a mandatory approach, or indeed to demonstrate that a consistent voluntary approach might deliver as well as a mandatory approach. Option 3 was the preferred option as it was felt that more evidence was required on the costs and benefits of biodiversity offsetting before developing a national policy

(continued)

Development	Date(s)	Description
National Ecosystem Assessment (NEA)	June 2011	Across the UK, a National Ecosystem Assessment (NEA) completed in 2011 concluded that land use change is the main driver of the fragmentation trend observed in English nature, and responsible for threatening more than 65% of farmland species (UK NEA 2011)
Natural Environment White Paper (NEWP) published	June 2011	The NEWP (HM Government 2011b) was the UK Government's first national environment White Paper for 20 years (DEFRA 2011c). It included a commitment to improve the environment for the next generation and over 90 commitments that would support this commitment. One of the commitments was to "*establish a new voluntary approach to biodiversity offsetting and will test this using a number of pilot areas*". This was a direct translation of the results from the Options Stage Impact Assessment (HM Government 2011a)
Pilot phase and evaluation announced	April 2012–2014	As set out in the Natural Environment White Paper DEFRA requested Local Planning Authorities (LPAs) to volunteer to join the biodiversity offsetting pilot phase. LPAs were not given any financial resource to join but they were provided with a set of materials developed by DEFRA and Natural England and received some support from Natural England staff The objective of the pilot phase was to: "develop a body of information and evidence that the Government can use to decide whether to support greater use of biodiversity offsetting, and if so, how to use it most effectively" (DEFRA 2011b) Six local councils participated as testing grounds for the policy's implementation in Devon, Doncaster, Essex, Greater Norwich, Nottinghamshire and Warwickshire, Coventry and Solihull. Other local authorities and/or organizations that expressed an interest to participate were attributed a complementary status by DEFRA, thus contributing their experience and results to build a strong evidence base (DEFRA 2013a, b) Alongside the pilot phase an evaluation project was commissioned (DEFRA 2011b)

(continued)

Development	Date(s)	Description
Owen Paterson Secretary of State	September 2012–July 2014	Owen Paterson was the Secretary of State for the Environment, Food and Rural Affairs from September 2012 to July 2014. Over that time he was a vociferous supporter of biodiversity offsetting, promoting it within articles, speeches, holding a summit with key stakeholders and travelling to Australia to learn from experience there. Paterson's enthusiasm stemmed from his view that biodiversity offsetting was a "*chance to improve the environment and grow the economy*" (Paterson 2013) Paterson suggested that national parks, ancient woodlands and Great Crested Newts could be included within England's biodiversity offsetting scheme (Hope 2013; Pitel et al. 2013). A number of commentators and environmentalists responded critically to Paterson's comments as recognised by Mr. Paterson himself (Johnston 2013; Monbiot 2013; Atkinson 2014; Swinford 2014)
Ecosystem Markets Task Force (EMTF)	February 2013	The EMTF was a commitment of the Natural Environment White Paper with the objectives of exploring opportunities for UK businesses to protect and/or value nature. The final report included a section on biodiversity offsetting and made the case that a mandatory system would have significant benefits for the economy generally, and developers specifically, whilst realising a substantial revenue stream for conservation activities (EMTF 2013). The report was cited by DEFRA within the Green Paper (DEFRA 2013a)
Green paper published	September 2013	In 2013 DEFRA published a Green Paper—a consultation document—on biodiversity offsetting policy (DEFRA 2013a). The document set out the case for biodiversity offsetting and requested evidence and opinions on the validity of this evidence and on a set of policy options. These options were: fully permissive (i.e. voluntary); partially permissive (where developers are obliged to use the metric but there is no requirement as to what happens subsequently); uniform approach (mandatory) based on a threshold of impact; and, an indirect approach whereby money to compensate for biodiversity loss is pooled with other funds The document presented options on a range of

(continued)

Development	Date(s)	Description
		other elements of a biodiversity offsetting system; such as the form of the metric; proximity between impact and offset; robustness of assessment; and sorts of habitats that can be offset The document includes a "*preference for giving developers the choice to use [biodiversity] offsetting*" (p. 5) but does not include a specific policy proposal Sullivan and Hannis (2015) provide an analysis of the consultation responses
Environmental Audit Committee of Biodiversity Offsetting	October 2013	After the Government had published the Green Paper, the Environmental Audit Committee (EAC) of the UK Houses of Common—which consists of Members of Parliament—opened an inquiry into biodiversity offsetting. The EAC collected evidence from a range of sources and produced a report including non-binding recommendations. DEFRA is required to provide a response to the EAC but nothing more The EAC's report reiterated the importance of the mitigation hierarchy, whilst suggesting that the metric used within the pilots is too simple; some habitats should be excluded; loss of access for local groups should be accounted for; and that protocols should be put in place to ensure local authorities are able to fully recover the costs of any offsets The EAC suggested that DEFRA should wait until the results of the evaluation of the biodiversity offsetting pilot phase were released before making any decision (Government had indicated that it was likely to pre-empt the evaluation and push forward with new policy) (Environmental Audit Committee 2013). The evaluation was due to finish seven months after the publication of the EAC's report. In their response DEFRA agreed to await the findings of the evaluation (Environmental Audit Committee 2014a)
Environmental Audit Committee of High Speed 2 (HS2)	April 2014	The EAC also reported in April 2014 on HS2 and the Environment (a new high-speed rail line between London and Birmingham, and then to Manchester and Leeds). The study included how biodiversity offsetting might be used to offset the significant impact of the new rail line on a large number of ancient woodlands and wildlife sites. The EAC highlighted that what was proposed by HS2 Ltd. was not strictly biodiversity offsetting, as defined by DEFRA, since it was all to take place within

(continued)

Development	Date(s)	Description
		the boundary of the route defined by the parliamentary Hybrid Bill process (Parliament has no remit to consent outside that area). The EAC highlighted especially that as a consequence HS2 Ltd's biodiversity offsetting proposals could never meet the 'no net loss' objective let alone net gain. This also raised concerns about the potential misuse of offsetting, something NGOs have become increasingly concerned about (Environmental Audit Committee 2014b)
Evaluation of pilot phase published	June 2014	The evaluation of the biodiversity offsetting pilot programme was a 2-year study undertaken with eight pilot areas. Evidence from the pilot programme suggested that whilst biodiversity offsetting has the potential to deliver improvements in biodiversity outcomes, there are barriers to its delivery. These included resource constraints, immaturity of the offset market and a lack of perceived support in planning policy. The report concluded that it will require additional resources and ecological expertise in local authorities to deliver a biodiversity offsetting policy. In general stakeholders considered the biodiversity offsetting metric to be an efficient and transparent tool to quantify and communicate the impact of a development on habitats. However, due to the fact that, in cases, the use of the metric introduced increased overall costs for developers, requirements for biodiversity compensation were often challenged Suggestions for improvement included the mandatory use of an enhanced biodiversity offsetting metric within planning applications, development of technical guidance for practitioners, strengthening of local policies on biodiversity loss and compensation and further options to improve the process and build the evidence base (Baker et al. 2014a)

To date DEFRA has been responsible for developing the policy and methodologies for biodiversity offsetting even though a different Government Department—the Department for Communities and Local Government (DCLG)—is responsible for the planning system and related guidance. It is understood that DCLG and the UK's Treasury are concerned that biodiversity offsetting may increase the cost of development – particularly of housing which is a political priority. One of the initial hopes of DEFRA was that "*[biodiversity offsetting] can be done without imposing new burdens on developers, and hope that a consistent approach to offsetting can make negotiations simpler and more straightforward for all involved"*. [page 6, DEFRA (2011a, b, c)] and this remains a priority. It is expected that until evidence emerges about the benefits of biodiversity offsetting and/or there is a change in the UK Government's priorities then the voluntary system currently in place will remain.

Studies on the voluntary approach suggest that demand for offsets "*is unlikely to be sufficient to support a viable habitat banking system*" (ten Kate et al. 2010) under current law and regulations (Treweek et al. 2009). Lessons from the international experience concur, revealing that in countries where biodiversity offsets were a legal requirement the policy was successful in generating benefits for the participants and the environment faster, easier and at a greater scale (Conway et al. 2013; Tucker et al. 2014). Other chapters in this book describe the experiences of mandatory systems in Austria, France or Germany. However, care is needed in interpreting how other countries' experience translates to the UK since the context (e.g. scale, planning system, local authorities and civil society is very different.

1.2.2 Principles for Biodiversity Offsetting

Making Space for Nature Review was an independent study commissioned by DEFRA (Lawton et al. 2010) and led by Professor Sir John Lawton. The review looked at England's network of wildlife sites and made a series of recommendations about how to improve their extent, quality and connectivity. The review considered the potential of biodiversity offsetting and recognised there were opportunities and risks. It proposed a set of principles for biodiversity offsetting that were intended to reduce the risks whilst maximising the potential benefits. These principles were largely taken up by DEFRA, which published a document on the *Guiding principles for biodiversity offsetting* (DEFRA 2011a) that includes the following principles:

"Biodiversity offsetting should . . .

1. Not change existing levels of protection for biodiversity
2. Deliver real benefits for biodiversity by:
    - (a) seeking to improve the effectiveness of managing compensation for biodiversity loss
    - (b) expanding and restoring habitats, not merely protecting the extent and condition of what is already there
    - (c) using offsets to contribute to enhancing England's ecological network by creating more, bigger, better and joined areas for biodiversity (as discussed in Making Space for Nature)
    - (d) providing additionality; not being used to deliver something that would have happened anyway
    - (e) creating habitat which lasts in perpetuity
    - (f) being at the bottom of the mitigation hierarchy, and requiring avoidance and mitigation of impacts to take place first

3. Be managed at the local level as far as possible:
   (a) within national priorities for managing England's biodiversity
   (b) within a standard framework, which provides a level of consistency for all involved
   (c) through partnerships at a level that makes sense spatially, such as county level, catchment or natural area
   (d) with the right level of national support and guidance to build capacity where it is needed
   (e) involving local communities
4. Be as simple and straightforward as possible, for developers, local authorities and others
5. Be transparent, giving clarity on how the offset calculations are derived and allowing people to see how offset resources are being used
6. Be good value for money"

In addition to these high-level principles a number of other principles are observed within the metric, these are set out in Sect. 1.2.3.

These principles were included in the pilot phase and formed part of the subsequent evaluation. Section 3 describes to what extent and in what ways these principles were implemented in practice.

1.2.3 Development of the Metric

Since 2009, in consultation with stakeholders, DEFRA and Natural England have been developing a metric to facilitate the quantification of biodiversity impacts. This has been an iterative process supported by reviews of international practice, consultations with conservation experts and piloting the metric with a number of local authorities. Along with the metric, Technical Guidance and general guiding materials have been produced to facilitate practitioners and stakeholders in managing and delivering biodiversity offsetting in a local context.

All of the matrices developed have used habitats "*as the primary metric*" (Treweek et al. 2009, p. 117). It was felt to be the most suitable unit "*conceptually*" (ibid) and practically as the UK Biodiversity Action Plan (BAP) provided an established list of important habitats, termed 'Priority Habitats'. According to guidance, biodiversity offsetting, and therefore the metric, would not apply and should not be used to weaken existing levels of legal protection for designated sites, such as SSSIs[5] or high value habitats, such as ancient woodlands, which cannot be replaced or re-created in reasonable time. In theory, this means that where a SSSI or ancient woodland is to be affected by a development, the planning authorities cannot take account of any compensatory works when making their decision as to whether

[5]Sites of Special Scientific Interest.

the planning application will have a significant impact. There are no examples of this occurring in practice.

DEFRA's initial work on the metric proposed that habitat parcels be defined based on their area and the types of habitats they included as well as the condition of these habitats. This approach has been sustained in the subsequent versions of the metric.

There are two stages of applying the metric; the first is calculating the impact of a development and the second is calculating the scale and type of offset required.

In the first stage, the metric uses the following approach: **habitat area × quality × condition.**

The metric defines habitat quality in intrinsic terms using one of three bands of 'distinctiveness'. Distinctiveness is a proxy for the biodiversity importance of the habitat. The number of definition of the bands were set through a series of consultations with experts and policy makers which took account of parameters such as species richness, diversity, rarity and whether a habitat supports species rarely found in others. Reflecting the fact that the metric is intended to support existing policy, the habitats listed as Priority Habitats in the BAP make up the top band and within the metric are equal to a score of 6 (High distinctiveness). Other semi-natural habitats are in the second band and score 4 (Medium distinctiveness). Artificial habitats, such as intensive farmland, are in the lowest band and score 2 (Low distinctiveness). Depending on local circumstances and making an assessment on a case by case basis, some authorities may choose to modify the distinctiveness assessments to take account of local contexts (e.g. a habitat of moderate distinctiveness is particularly rare in the local area) (Rayment et al. 2014).

The metric includes three levels for habitat condition: good, moderate and poor (3, 2 and 1 respectively). At the time of writing there is limited relevant guidance available on how to undertake a condition assessment. During the pilot phase the Farm Environment Plan Handbook (Natural England 2013) was used but this was known to be only partly suitable as it considered farmed habitats only.

The area, quality and condition of habitat are calculated and expressed as biodiversity units using the matrix in Table 1.

These scores are multiplied by the area of habitat to calculate the total number of biodiversity units lost (which are equivalent to the habitat hectare units commonly used in other metrics related to offsetting). Table 2 shows an example taken from DEFRA Affairs (2012b).

The offset element of the metric uses the same basic structure but also considers delivery and spatial risk and a temporal multiplier. These are defined as (Depa DEFRA 2012a):

- Delivery risk—"The risks associated with the actual delivery of the offset due to, for instance, uncertainty in the effectiveness of restoration or habitat creation/ management techniques." (DEFRA 2012a). The scale of delivery risks runs from 10 to 1 meaning that a multiplier of 10 is applied to the number of conservation

Table 1 Use of the habitat-based matrix to calculate number of biodiversity units

		Distinctiveness		
		Low (2)	Medium (4)	High (6)
Condition	Good (3)	6	12	18
	Moderate (2)	4	8	12
	Poor (1)	2	4	6

Source: DEFRA (2012a)

Table 2 Use of the habitat-based matrix to calculate number of biodiversity units

Habitat	Distinctiveness	Condition	Hectares	Number of units
Lowland meadow	6	2	6	(6 × 2 × 6) 72 biodiversity units lost

credits that are required to offset the loss of habitats that are felt to be very difficult to create or restore. This was based on an assessment of international practice (Treweek et al. 2009). Depending on the difficulty to restore a site the following multipliers are applied:

- Very High = 10
- High = 3
- Medium = 1.5
- Low = 1

- Spatial risks—"These reflect ecological risks deriving from the change in location of the habitat or resource. For example, it may be that recreating a type of habitat in a new location may reduce its biodiversity value" (DEFRA 2012a). During the pilot phase, this was calculated on the basis of the pilots offsetting strategies, so if an offset site was located in an area identified by a pilot offset strategy then a multiplier of 1 was used, if it was not then a multiplier of 3 was applied. In cases where the offset was located in an area where it was buffering, linking, restoring or expanding a habitat outside an area identified in the offsetting strategy, a multiplier of 2 was used.
- The temporal multiplier calculates a discount rate (3.5% per year) to the amount of time it will take the offset habitat to achieve 'target condition' i.e. be of comparable quality to the lost habitat. For instance, if the years to target condition are 10 years a multiplier of 1.35 is applied to the number of conservation credits required. The multiplier is capped to a maximum of 3 (which is only equivalent to 32 years, and therefore applies to most habitat types).

Annex 1 of Rayment et al. (2014) and DEFRA (2012a) provide more detail on the metric and its application within England.

The metric includes a principle referred to as 'trading up'. This means that if a highly distinctive habitat is to be lost then it must be replaced with the same habitat type or if necessary another highly distinctive habitat. If the habitat is of medium or low distinctiveness the offset must be medium or, preferably, highly distinctive.

Therefore, while biodiversity offsetting should ideally be like-for-like, for medium and low distinctive habitats trading up can occur if a suitable habitat of the same type is not available in the offsetting area.

Government consultations (DEFRA 2013a, b; Environmental Audit Committee 2013) have identified concerns around the 'simplistic' nature of the metric and the fact it does not directly account for affected species. Evidence from the pilot phase is not consistent on this subject, with some viewing the relatively simple mechanics of the metric as a major strength and others feeling it underestimates the complexity of biodiversity and therefore underestimates any related loss (Baker et al. 2014a).

Overall, developers seem to welcome this element of the metric as it is "*easy to interpret and does not require lengthy and costly surveys*" (Environmental Audit Committee 2013). The same simplicity is criticised by others notably conservation organisations, "*in particular regarding the low number of distinctiveness and condition categories, its incompleteness regarding assessment of some biodiversity attributes (such as ecological networks) and the need for value judgements to be made*" (Rayment et al. 2014). With reference to species, and although the biodiversity offsetting pilots in the UK have not directly incorporated species in the metric, one local authority in England has been developing and using a metric that accounts for species for a number of years. Somerset County Council initially used their metric to support assessments of impacts on European Protected Species before adapting it to include other designations and non-designated areas too (Baker et al. 2014c). Although the local authority has promoted the metric and it has been used on a large number of applications, there are concerns that the quantity and quality of species data required is not available across the rest of England and its wider adoption is not feasible without an improvement in ecological data (Baker et al. 2014a). Another limitation of Somerset's approach is that different species may be important in different areas, requiring a significant degree of local interpretation, which would make the development and implementation of the metric problematic.

To date DEFRA and Natural England have been responsible for developing and hosting the metric. As local authorities and environmental ecologists develop capacity and expertise around biodiversity offsetting, multiple matrices—all based on the architecture developed by DEFRA—will emerge. Within the current voluntary approach this presents challenges and opportunities to developers and offset providers who operate across multiple areas and local authorities have already been challenged on their interpretation of the metric (Baker et al. 2014a). If the presentation and mechanics of the metric continue to be adapted it is possible that DEFRA may be asked to provide additional guidance. This is potentially problematic as recent UK Governments have promoted local decision making (referred to as localism) whilst reducing the amount of guidance produced by Government Departments and Agencies (as part of general reduction in the scale of Government).

2 Practice in England

2.1 *Biodiversity Offsetting Pilot Phase*

In July 2011, DEFRA requested Expressions of Interest (EoI) from Local Planning Authorities (LPAs) to join a scheme to test a form of voluntary biodiversity offsetting. The pilot phase was promoted on the basis of the 2011 Making Space for Nature Review (DEFRA 2011a) and subsequent Natural Environment White Paper (NEWP) both of which set out DEFRA's intention to test a voluntary approach to biodiversity offsetting. In this instance, a voluntary approach meant no change to planning policy and developers were not required to use the metric or to use biodiversity offsetting to compensate for residual loss. It was for the local authorities, within existing policy, to promote biodiversity offsetting as they saw fit.

The pilots were selected in September 2011 with the pilot scheme starting in April 2012 and finishing in March 2014. The objective of the pilots was "*to develop a body of information and evidence to inform a future decision about whether to use biodiversity offsetting across England*" (Department for Environment, Food and Rural Affairs 2011b). During the pilot phase, Natural England provided each of the pilots with up to 0.5 full time equivalents (FTE) of Adviser time. The Advisers worked with the pilots to advise on the development of their biodiversity offsetting strategy, the assessment of biodiversity impacts and use of the metric. It was also intended that the Advisers would accredit offset providers and review the capability and viability of any specific offset proposals.

As set out in Sect. 1.2.3 DEFRA and Natural England had developed a metric and Technical Guidance to facilitate the measurement of biodiversity impacts. These materials provided guidance and some structure for how the pilots should manage and deliver biodiversity offsetting within their local context. Less technical guidance was also produced for the key parties in the pilots: local authorities, developers and offset providers.

It was hoped that the pilots would be able to test and provide learning on all aspects of biodiversity offsetting including providing a clearer sense of the costs and benefits for nature and developers. In reality, the pilot programme was not able to advance as far as initially hoped and by the end of the 2 years only one of the pilots had a formal agreement to offset a development that had received planning permission.

Despite this there were some achievements, for instance all of the pilots developed some form of biodiversity offsetting 'strategy'; sought to incentivise biodiversity offsetting generally and within specific developments; and raised capacity across local authorities and other groups—notably ecological consultants. While few actual 'offset' projects were agreed or delivered (in terms of offsite habitat creation on land owned by a third party and funded by developer contributions), the metric was widely used. In fact, all but one of the pilots (Greater Norwich) used the metric on live planning applications, but only two [Essex and the Coventry, Solihull, Warwickshire Association of Planning Officers (CSWAPO)] progressed to a complete biodiversity offsetting project. This is where an offset receptor site has been identified and is expected to be implemented, although only one was supported by a legal agreement.

The Environment Bank[6] is currently the only biodiversity offsetting broker in the UK taking on the role of an intermediary between offset providers and buyers by developing an online exchange platform. Its role lies in connecting participants in the market and facilitating the recording, credit calculation and trading of biodiversity offsets. Environment Bank provided two of the pilots (Essex and CSWAPO) with resource to support their work and it was in these two pilots where some offsets were agreed—at least in principle.

Within Essex and CSWAPO, four organisations were interested in providing offset sites on their land to secure additional funding. Two of these were local authorities who were hoping to secure funding to change management regimes and increase the biodiversity value of land for which they were responsible (including amenity grassland, parks and nature areas). This was in part a response to their budgets decreasing. The potential providers from local authorities did note that because of the broad remit of their organisations and the need to demonstrate value for money, any offset site would have to provide multiple benefits, including education, volunteering and recreation. It was felt that although this might reduce the biodiversity value of any offsets it was a manageable risk. One of the other potential providers was a conservation group that was looking to improve existing reserves or, if sufficient funds were made available, to acquire new reserves. The conservation group had a number of concerns around biodiversity offsetting and in particular the need to ensure that offsets were not being secured from developments that were causing the loss of significant quantity or quality of biodiversity. This meant that they wanted to be involved in the original planning application to ensure that the principles for biodiversity offsetting were adhered to. The other potential provider was a commercial business that worked with landowners and farmers within the agricultural supply chain. They were interested in brokering relationships between landowners and the Environment Bank. All the potential providers had concerns about the requirement for long term management, which tended to be 20–25 years, and the reduction in flexibility this meant. They were also reticent to develop management plans with no assurance of receiving funding; as planning permission for the related development may be refused.

Despite not fully meeting DEFRA's expectations, the pilot scheme has provided valuable information for policy makers in terms of testing the processes whilst highlighting gaps in knowledge and/or capacity that need to be satisfied for a successful system to be implemented.

The pilots have also demonstrated that the current lack of demand for offsets, linking back to the voluntary nature of the policy, does not encourage the supply of land for offsets, creating an overall lack of liquidity in the offset market. In practice, this has meant that developers have difficulty in finding a suitable offset which

[6]http://www.environmentbank.com/

results in increased transaction costs and a poorer quality of offset site. The pilot phase has demonstrated that an effective system that delivers significant net benefits for biodiversity requires a critical mass of offsets so that the required processes and systems can be put in place and appropriately resourced (DEFRA 2013a, b).

Ecosystem Service Offsetting?

One of the aims of the North Devon pilot was to explore how the principles and mechanisms of biodiversity offsetting might address losses in ecosystem services. The pilots developed a process but were not able to test it. In summary, the process included using land cover maps and assigning ecosystem services to different habitat types [informed by the UK's National Ecosystem Assessment (2011)]. The expectation was that losses of ecosystem services from development would be replaced on a like-for-like basis e.g. if habitats that provided flood attenuation ecosystem services were lost they would need to be replaced by habitats with a similar ecosystem service function.

In parallel, a number of local authorities involved in or aware of the pilots voiced concerns that local communities could lose green spaces to development and have it replaced by inaccessible or remote offset sites. This led to discussions between DEFRA, Natural England and the pilots about incorporating ecosystem services into the metric to account for, and therefore reduce the loss of cultural (or other) ecosystem services.

It was felt that at the time the metric should focus on biodiversity alone and that non-biodiversity issues would be dealt through existing planning processes, for instance, many developments are required to consider their impact on flood risk/recreation opportunities and put in place appropriate mitigation or compensation. Attempting to do this via the metric was considered technically challenging and inconsistent with how planning decisions were made (Baker et al. 2014c).

2.2 *Lessons from the Biodiversity Offsetting Pilot Phase*

The following lessons for practice were identified from the biodiversity offsetting pilot phase (adapted from Baker et al. 2014a):

- The metric is an effective biodiversity impact tool
  - When the metric is used, it suggests that the vast majority of planning permissions will result in the loss of some biodiversity. This is because most applications involve some land take and the metric ascribes a biodiversity value to even low quality and low distinctiveness land.

  - Using the metric to calculate biodiversity loss is an effective way of increasing the visibility of biodiversity loss—in particular the ongoing failure to implement net-gain policy within planning applications.
  - Within a voluntary system, the main value of the metric is increasing onsite mitigation and compensation. Requiring offsite compensation is felt to be too challenging within current policy—although the Coventry, Solihull and Warwickshire Association of Planning Officers (CSWAPO) pilot did manage to address this (see Sect. 11.2.3).
  - Using the metric to quantify the biodiversity impact of a planning application will likely increase the costs to the developer. This is because, compared to current practice, most applications do not fully identify or account for biodiversity losses.

- Institutions; resources and roles

  - Local authorities will need additional ecological expertise and capacity if biodiversity offsetting is to be reliably effective and become more common/ mandatory.
  - There are tensions between Natural England's policy of focusing their increasingly limited resource on high risk cases (those potentially affecting statutorily protected sites) and biodiversity offsetting having been designed to deal with lower levels of biodiversity loss. This meant that the pilots were rarely able to work with local Natural England officers to resolve emerging problems. At a national level, if biodiversity offsetting is to be viable the role of the regulator must be clarified and appropriately resourced.

- Planning policy and implementation

  - Current policy states that designated sites such as SSSIs and ancient woodland cannot be offset but these sites continue to be damaged and lost to development.
  - Unless standards—including their enforcement—for onsite mitigation and compensation are raised it is likely that much of the onsite activity that is being stimulated through the use of the metric will be on paper only.
  - Competition for land in England is intense as is any compensation money negotiated by local authorities with developers through s106 agreements. This means that if developers are required to pay for biodiversity loss (where they have not in the past) they are likely to seek to negotiate trade-offs with other public benefits to be delivered, e.g. social housing provision. There is an expectation that within a mandatory system these losses will be capitalised from the initial cost of land acquisition. This was not tested within the pilots.
  - Maintaining any offset in 'perpetuity' is one of the principles developed by DEFRA. In practice perpetuity is likely to be interpreted as between 20 and 30 years unless clearer guidance is provided.
  - The ability of biodiversity offsetting to 'streamline' planning applications is quite limited. This is because, with the exception of a few instances, issues relating to compensating for the impact on non-designated biodiversity are not

responsible for holding up planning applications. This is evidenced by the fact that as few as 0.1% of all planning applications are required to deliver any offsite compensation. However, there is some evidence that the biodiversity aspects of planning applications can be agreed more quickly and efficiently as the metric provides a common basis for discussions around mitigation and compensation.

- Public perception
  - Many members of conservation groups are philosophically opposed to biodiversity offsetting. As the profile of biodiversity offsetting increased during the pilot phase it was apparent that this had reduced the willingness of conservation groups to engage and to proffer offset sites.
  - Local authorities noted that a 'red line' for them was that any offset must be provided for within the same local authority area and ideally as close as possible to the site of the loss. This was to account for concerns that offsetting would mean habitats and developer contributions leaving the local authority. This problem was cited by some local authorities as one of the reasons they did not join the pilot phase (Baker et al. 2014a).
- Limits of voluntary
  - A market for offsets in England is unlikely to develop whilst a voluntary approach is in place. If this changes it will also take some time to develop a market as there are many barriers for potential offset providers. Most notably the lack of clarity about the long-term policy support for biodiversity offsetting and the expectations of local authorities for offsets to be located as locally as possible so as not to lose biodiversity (and/or money) from their administrative area.

2.3 Case Studies

This section presents three case studies that demonstrate different elements of applying biodiversity offsetting in England (Baker et al. 2014c). The cases are:

- The Coventry, Solihull and Warwickshire Association of Planning Officers (CSWAPO) biodiversity offsetting pilot.
- Planning application within Essex and concerns over the mitigation hierarchy.
- Example of voluntary biodiversity offsetting from Nottinghamshire.

2.3.1 The Coventry, Solihull and Warwickshire Association of Planning Officers (CSWAPO) Biodiversity Offsetting Pilot

The Coventry, Solihull and Warwickshire Association of Planning Officers (CSWAPO) led one of the six biodiversity offsetting pilots that were set up by DEFRA. CSWAPO is notable for two reasons; the extent to which they used the metric, and their strong interpretation of 'significant' loss of biodiversity which allowed them to implement a mandatory form of biodiversity offsetting.

Within the pilot the lead organisation was Warwickshire County Council but the pilot was jointly hosted and managed by CSWAPO[7] and is commonly referred to as the CSWAPO pilot. The local authorities represented by CSWAPO were notable as, in addition to high quality biodiversity data and mapping, they had a relatively large team of ecologists. This capacity was an important reason why, by the end of the 2 years of the pilot phase, the pilot hosts had developed a process where the metric was used on almost all of the planning applications that were being received. In total, the metric was applied to 63 applications within March 2012–2014. None of the other pilots applied the metric to more than 10 applications.

After the metric had been applied the pilot worked with local authority ecologists to consider how the impact of the development could be reduced. This included reformulating the development to avoid more important areas of biodiversity and looking for onsite mitigation and compensation. If after this process the development was considered likely to result in any 'significant loss', as determined by the metric, then offsite compensation would be required. Significant loss was determined to be where the metric showed that more than one biodiversity unit was to be lost and/or the habitat was considered to be locally or nationally important as defined by local planning policy and the UK Biodiversity Action Plan.

Where significant loss was identified the local authority undertook discussions around how the developers might reduce the scale of the loss through appropriate on-site compensation where possible. In those applications where sufficient onsite compensation was not possible the local authority would recommend biodiversity offsetting as an appropriate mechanism to deliver offsite compensation, to be secured by s106 agreements. The local authority supported this recommendation by referring to the requirement to deliver biodiversity 'net-gain' as set out in national planning guidance and a broader interpretation of 'significant loss' to include any development which was causing the loss of more than one biodiversity unit.

Compared to the other pilots this process was very successful in getting increased levels of mitigation and compensation. There were multiple instances where applications had been adapted to reduce their impact on biodiversity and agreements were made to use biodiversity offsetting in a number of instances.[8]

---

[7]CSWAPO was formed of: Warwickshire County Council; Coventry City Council; Solihull Metropolitan Borough Council; North Warwickshire Borough Council; Nuneaton and Bedworth Borough Council; Rugby Borough Council; Warwick District Council; and Stratford District Council.

[8]Due to the long timescales involved with planning applications only one biodiversity offset had been agreed and this had not been formalised before the end of the pilot phase.

2.3.2 Planning Application Within Essex and Concerns Over the Mitigation Hierarchy

One of the concerns about biodiversity offsetting has been that by improving the acceptability and quality of offsite compensation developers and planning authorities would be more likely to develop existing habitats as they could argue that their development will result in a net-gain. To date, there is limited evidence that this might happen, across the whole planning system, but there were indications within the pilots that that it might occur in some instances.

One relevant example occurred in the Essex biodiversity offsetting pilot where an application was made for approximately 50 residential units on a site of species-rich unimproved grassland. This habitat was considered to be of sufficient quality to meet Local Wildlife Site designation (which is a non-legal form of protection, recognised through the planning system). The original application included no offsite compensation and would result in loss of much of the grassland. The local authority felt that the application included insufficient mitigation and compensation measures and was rejected; other non-biodiversity reasons contributed to the initial refusal.

After the initial refusal, the developers worked with a third party to develop a proposal for offsite compensation. This was included in the appeal (a legal process where the planning application is effectively resubmitted). At appeal the developers argued that, as no management was planned or funded on the affected grassland, it would deteriorate over time. In comparison, it was argued, the proposed offsite compensation would deliver long term biodiversity benefits by dint of the commitment to long term monitoring and management included within the offset agreement. The planning inspector agreed with the developer and the permission was approved.

The revised application retained few of the original onsite features. It is arguable that in this instance there was an over-reliance on biodiversity offsetting to meet the requirement of net-gain. The success of this application at appeal may create precedent for future applications and there is a risk that without additional guidance the mitigation hierarchy may be undermined by biodiversity offsetting providing a validated route for compensation.

2.3.3 Example of Voluntary Biodiversity Offsetting from Nottinghamshire

A large (120 ha) mixed use development was proposed in the county of Nottinghamshire, which was one of the biodiversity offsetting pilots. Due to the scale of the application and the fact that the development site included grassland that was designated as a Local Wildlife Site the pilots looked to use biodiversity offsetting.

The original application included no onsite mitigation or compensation. Subsequently the applicants were required to undertake an EIA. During the initial stage of the EIA the local authority requested that the biodiversity offsetting metric be applied. The applicants were initially reluctant to use the metric, but the local authority insisted and indicated that using the metric would strengthen the application. Having used the metric a series of mitigation measures were identified to reduce the onsite biodiversity loss. The developers and local authority worked together to develop the onsite measures and the translocation of part of the grassland—no offsite compensation was agreed.

The local authority estimated that between 78 and 90% of the biodiversity loss had been reduced through the onsite proposals; assuming that the onsite measures are delivered. Although this does not represent net-gain, the local authority felt that this was the best solution it could achieve within existing planning policy and that the political pressure to promote development made a more robust interpretation of net-gain impossible to implement. Within this instance the metric was highlighted as central to substantially reducing the biodiversity loss. In particular the clarity and quantification provided by the metric provided a platform for the local authority to negotiate improved mitigation.

3 Policy and Practice in Devolved Administrations

The United Kingdom consists of three Devolved Administrations each of which is responsible for the majority of their own environmental policy. To date England has led on the development and testing of biodiversity offsetting with Wales, Scotland and Northern Ireland exhibiting varying levels of discussion and experience.

Wales

As with England, Wales has national planning policies which allow for offsite compensation where "*damage is unavoidable*" (Welsh Government 2014); this is supported by Technical Advice Note 5 (Welsh Government 2009) that includes additional detail on the nature and use of mitigation and compensation, but does not include any reference to biodiversity offsetting or any features unique to biodiversity offsetting such as metrics.

To date the Welsh Government has produced briefing papers on biodiversity offsetting and the Sustainability Committee of the 3rd Welsh Assembly (the Welsh parliament) undertook an inquiry into biodiversity that included discussions around the opportunities and threats of biodiversity offsetting (National Assembly for Wales 2014). As yet no specific policies or recommendations for future policies have emerged from these discussions.

Scotland

In June 2013, the Scottish Government published a consultation version of their Biodiversity Strategy which included a request for views on biodiversity offsetting (Scottish Government 2013). In response to the opinions received the Scottish Government published a response that said:

> A variety of concerns were raised about the potential for biodiversity offsetting, given the range of concerns raised, the Scottish Government does not plan to consider biodiversity offsetting further at this time but recognises that Scottish planning authorities have used

primary agreements to secure biodiversity actions to offset damage to a site caused by developments in particular cases (Scottish Government 2013)

Evidence suggests that local authorities have been exploring biodiversity offsetting, but on a voluntary basis (Edinburgh Centre for Carbon Innovation 2013). One notable example is the Scottish Borders Council (a local authority), that has used existing legal and planning systems to offset the loss of upland habitats from a number of onshore wind farms (see Tharme and Aikman (2012) for more).

Northern Ireland

As with England and the other Devolved Administrations, compensation for biodiversity loss is an established if ad-hoc part of planning policy in Northern Ireland. To date there have been no publicly available discussions or policy developments around biodiversity offsetting. The most recent publication of the Strategic Planning Policy Statement in 2015 (Department of the Environment Northern Ireland 2015) makes no reference to biodiversity offsetting.

4 Conclusions from a United Kingdom Practitioner's Perspective on an EU-wide No Net Loss and Offset Strategy

Though there is currently no mandatory framework in England and the UK Devolved Administrations, much of the necessary architecture for widespread biodiversity offsetting is in place. What is currently missing is a national decision to make biodiversity offsetting, or the use of the metric, mandatory. Part of the reluctance for any such decision is the lack of experience and evidence related to requiring, delivering and monitoring offsite compensation.

Ultimately, addressing the failures of the current system which delivers a net loss of biodiversity over time is not considered a political priority.

While a mandatory biodiversity offsetting system would make clear the insufficiency of current biodiversity mitigation and compensation, it would increase costs for developers at a time when the Government is looking to reduce these as far as possible. In addition, local authorities are likely to continue to want to see any offsets as locally as possible so as not to export benefits beyond their administrative area, which suppresses significantly the viability of a market in offsets. The current voluntary policy seems likely to continue, therefore, in the medium term even though there will be continued loss of biodiversity through development, as highlighted by the metric.

References

Atkinson S (2014) Enough is enough – protect Britain's ancient woods! The Ecologist. http://www.theecologist.org/campaigning/2316316/enough_is_enough_protect_britains_ancient_woods.html. Accessed 27 Oct 2015

Baker J, Sheate WR, Bennett T, Payne D, Tucker G, White O, Forrest S (2014a) Evaluation of the biodiversity offsetting pilot programme final report. DEFRA, London

Baker J, Sheate WR, Papadopoulou L, Bennett T, Payne D, Bull J (2014b) Indicative costs of current compensation arrangements for biodiversity loss: illustrative case studies. DEFRA, London

Baker J, Sheate WR, Bennett T, Payne D, White O, Forrest S (2014c) Evaluation of the biodiversity offsetting pilot programme, Pilot Reports, vol 2. DEFRA, London

Conservative Party (2010) Open source planning Green Paper. Policy Green Paper No. 14

Conway M, Rayment M, White A, Berman S (2013) Exploring potential demand for and supply of habitat banking in the EU and appropriate design elements for a habitat banking scheme. Final report submitted to DG environment. ICF GHK, London

Department for Communities and Local Government [DCLG] (2012) National planning policy framework. https://www.gov.uk/government/uploads/system/uploads/attachment_data/file/6077/2116950.pdf. Accessed 2 Nov 2015

Department for Communities and Local Government [DCLG] (2015) Planning practice guideline natural environment. https://consult.defra.gov.uk/biodiversity/biodiversity_offsetting/supporting_documents/20130903Biodiversity%20offsetting%20green%20paper.pdf, http://planningguidance.communities.gov.uk/about/. Accessed 27 Oct 2015

Department for Environment, Food and Rural Affairs [DEFRA] (2007) Conserving biodiversity—the UK approach. DEFRA, London

Department for Environment, Food and Rural Affairs [DEFRA] (2011a) Biodiversity offsetting: guiding principles for biodiversity offsetting. https://www.gov.uk/government/uploads/system/uploads/attachment_data/file/218681/110714offsetting-guiding-principles.pdf. Accessed 27 Oct 2015

Department for Environment, Food and Rural Affairs [DEFRA] (2011b) Evaluation of the biodiversity offsetting pilot phase – WC1051. http://randd.defra.gov.uk/Default.aspx?Menu=Menu&Module=More&Location=None&Completed=0&ProjectID=18229. Accessed 3 Nov 2015

Department for Environment, Food and Rural Affairs [DEFRA] (2011c) Natural environment. White paper. https://www.gov.uk/government/publications/natural-environment-white-paper-implementation-updates. Accessed 3 Nov 2015

Department for Environment, Food and Rural Affairs [DEFRA] (2012a) Technical paper: the metric for the biodiversity offsetting pilot in England. https://www.gov.uk/government/uploads/system/uploads/attachment_data/file/69531/pb13745-bio-technical-paper.pdf

Department for Environment, Food and Rural Affairs [DEFRA] (2012b) Biodiversity offsetting pilots: information note for local authorities. https://www.gov.uk/government/uploads/system/uploads/attachment_data/file/69529/pb13744-bio-local-authority-info-note.pdf. Accessed 28 Nov 2015

Department for Environment, Food and Rural Affairs [DEFRA] (2013a) Biodiversity offsetting in England. Green paper. https://consult.defra.gov.uk/biodiversity/biodiversity_offsetting/supporting_documents/20130903Biodiversity%20offsetting%20green%20paper.pdf. Accessed 27 Oct 2015

Department for Environment, Food and Rural Affairs [DEFRA] (2013b) Biodiversity offsetting in England. Green paper summary of responses. https://www.gov.uk/government/uploads/system/uploads/attachment_data/file/501240/biodiversity-offsetting-consult-sum-resp.pdf. Accessed 4 Oct 2016

Department of the Environment Northern Ireland [DoENI] (2015) Strategic planning policy statement for Northern Ireland (SPPS): planning for sustainable development (September 2015). http://www.planningni.gov.uk/SPPS. Accessed 4 Oct 2016

Drayson K, Thompson S (2013) Ecological mitigation measures in english environmental impact assessment. J Environ Manag 3:1–8

Ecosystem Markets Task Force [EMTF] (2013) Realising nature's value: the final report of the ecosystem markets task force. https://www.gov.uk/government/uploads/system/uploads/attachment_data/file/316101/Ecosystem-Markets-Task-Force-Final-Report-.pdf

Edinburgh Centre for Carbon Innovation [ECCI] (2013) Biodiversity offsetting and habitat banking for Scotland: notes from the workshop, Oct 2013. http://edinburghcentre.org/files/documents/SD_Biodiversity_Offsetting_and_Habitat_Banking_for_Scotland_FINAL_REPORT.pdf. Accessed 4 Oct 2016

Environmental Audit Committee [EAC] (2013) Biodiversity offsetting. Sixth Report of Session 2013–14. http://www.publications.parliament.uk/pa/cm201314/cmselect/cmenvaud/750/750.pdf. Accessed 4 Oct 2016

Environmental Audit Committee [EAC] (2014a) Ninth special report biodiversity offsetting: government response to the committee's sixth report of session 2013–14. http://www.publications.parliament.uk/pa/cm201314/cmselect/cmenvaud/1195/1195.pdf. Accessed 4 Oct 2016

Environmental Audit Committee [EAC] (2014b) HS2 and the environment. Thirteenth Report of Session 2013–14. http://www.publications.parliament.uk/pa/cm201314/cmselect/cmenvaud/1076/1076.pdf. Accessed 4 Oct 2016

Green Party (2015) Manifesto 2015 general election. https://www.greenparty.org.uk/we-stand-for/2015-manifesto.html. Accessed 4 Oct 2016

HM Government (2011a) Options stage impact assessment: offsetting the impact of development on biodiversity. https://www.gov.uk/government/uploads/system/uploads/attachment_data/file/218685/newp-ia-offsets-110607.pdf. Accessed 29 Oct 2015

HM Government (2011b) The natural choice: securing the value of nature. http://www.official-documents.gov.uk/document/cm80/8082/8082.pdf. Accessed 26 Oct 2015

Hope C (2013) Developers can build in national parks if they make up for the damage elsewhere, Owen Paterson to say. The Telegraph. http://www.telegraph.co.uk/news/earth/hands-off-our-land/10332079/Developers-could-build-in-national-parks-if-they-make-up-for-the-damage-elsewhere-Owen-Paterson-to-say.html. Accessed 16 Nov 2015

Institute for Environmental Management (IEMA) (2011) The state of environmental impact assessment practice in the UK [Online]. Available via https://transform.iema.net/article/state-eia-practice-uk

Johnston I (2013) 'Licence to trash nature': campaigners warn of new scheme that would allow house-building in biodiverse areas. The Independent. http://www.independent.co.uk/environment/nature/licence-to-trash-nature-campaigners-warn-of-new-scheme-that-would-allow-house-building-in-biodiverse-8919217.html. Accessed 29 Oct 2015

Lawton J, Brotherton PNM, Brown VK, Elphick C, Fitter AH, Forshaw J, Haddow RW, Hilborne S, Leafe RN, Mace GM, Southgate MP, Sutherland WJ, Tew TE, Varley J, Wynne GR (2010) Making space for nature: a review of England's wildlife sites and ecological network. http://archive.defra.gov.uk/environment/biodiversity/documents/201009space-for-nature.pdf. Accessed 27 Oct 2015

Milieu, IEEP and ICF (2016) Evaluation study to support the fitness check of the birds and habitats directives. Milieu Ltd, Institute for European Environmental Policy and the ICF International, Brussels

Monbiot G (2013) Biodiversity offsetting will unleash a new spirit of destruction on the land. The Guardian. http://www.theguardian.com/environment/georgemonbiot/2012/dec/07/biodiversity-offsetting-unleash-wildlife-destruction. Accessed 30 Oct 2015

National Assembly for Wales (2014) Sustainability committee, inquiry into biodiversity in wales. http://www.assembly.wales/Laid%20Documents/CR-LD8384%20-%20Sustainability%20Committee%20Inquiry%20into%20biodiversity%20in%20Wales-31012011-208859/cr-ld8384-e-English.pdf. Accessed 4 Oct 2016

Natural England (2013) Higher level stewardship: environmental stewardship handbook, 4th edn, January 2013. http://publications.naturalengland.org.uk/publication/2827091. Accessed 03 Jun 2016

Newey G (2012) Nurturing nature: policy to protect and improve biodiversity. Policy Exchange, London

Office of National Statistics (ONS) (2012) Planning applications: April to June 2012 [Online]. Available via http://webarchive.nationalarchives.gov.uk/20121102211427/http://www.communities.gov.uk/publications/corporate/statistics/planningapplicationsq22012
Parliamentary Office of Science and Technology [POST] (2013) Biodiversity and planning decisions parliament. London
Paterson O (2013) 'Biodiversity offsetting – a chance to improve the environment and grow the economy'. Business Green. http://www.businessgreen.com/bg/opinion/2293143/biodiversity-offsetting-a-chance-to-improve-the-environment-and-grow-the-economy. Accessed 29 Oct 2015
Pitel L, Thomson A, Sylvester R (2013) 'Newt credits' could settle environmental disputes. The Times. http://www.thetimes.co.uk/tto/news/uk/article3655504.ece. Accessed 16 Nov 2015
Rayment, M., White, A., Dickie, I. and Pieterse, M. (2011) Costing potential actions to offset the impacts of development on biodiversity. Final report. Report for DEFRA submitted by GHK in association with Eftec, GHK Consulting, London
Rayment M, Haines R, McNeil D, Conway M, Tucker G and Underwood E (2014) Study on specific design elements of biodiversity offsets: biodiversity metrics and mechanisms for securing long-term conservation benefits. A Report to the European Commission, DG Environment
Scottish Government (2013) 2020 challenge for Scotland's biodiversity – a strategy for the conservation and enhancement of biodiversity in Scotland, June 2013
Sullivan S, Hannis M (2015) Nets and frames, losses and gains: value struggles in engagements with biodiversity offsetting policy in England. Ecosyst Serv 15:162–173
Swinford S (2014) 'Owen Paterson: I've planted trees so can't be accused of 'trashing' woodlands'. The Telegraph. http://www.telegraph.co.uk/news/earth/environment/10560790/Owen-Paterson-Ive-planted-trees-so-cant-be-accused-of-trashing-woodlands.html. Accessed 29 Oct 2015
ten Kate K, Treweek J, Ekstrom J (2010). The use of market-based instruments for biodiversity protection – the case of habitat banking. Technical Report for European Commission DG Environment. EFTEC, London
Tharme A, Aikman I (2012) Biodiversity offset schemes in the borders. http://bit.ly/1M1TXT9. Accessed 29 Oct 2015
Treweek J, ten Kate K, Butcher B, Venn O, Garland L, Wells M, Moran D, Thompson S (2009) Scoping study for the design and use of biodiversity offsets in an english context. https://www.gov.uk/government/uploads/system/uploads/attachment_data/file/218689/BiodiversityOffsets12May2009.pdf. Accessed 5 Nov 2015
Tucker GM, Allen B, Conway M, Dickie I, Hart K, Rayment M, Schulp CJE (2014) Policy options for an EU no net loss initiative. Report to the European Commission (with Annexes). Institute for European Environmental Policy, London
United Kingdom National Ecosystem Assessment [UK NEA] (2011) National ecosystem assessment. http://uknea.unep-wcmc.org/. Accessed 4 Oct 2016
Welsh Government (2009) Technical Advice Note 5, Nature Conservation and Planning. September 2009 para 4.6.1, 4.7.1, 4.7.2
Welsh Government (2014) Planning policy wales: conserving and improving natural heritage and the coast. February 2014 para 5.2.8

Other EU Member States

Wolfgang Wende

In addition to the approaches discussed in detail in previous chapters, there exist other concepts of biodiversity offsets in various EU member states. Clearly, not all of these approaches can be scrutinized in this book and, indeed, not all approaches are well known. For these reasons we hope that the reader appreciates the impossibility of providing a detailed study of all European models, and that hence some approaches are necessarily absent from discussion. However, the range of presented models from diverse countries confirms both the widespread uptake of methods and instruments of biodiversity offsetting as well as the existence of an intensive scientific and political debate on this topic. These coalescing methods and instruments could, in fact, serve as the basis for a future set of European guidelines on biodiversity offsets. Further, the various national approaches can help show how best to implement the requirements of the EU Biodiversity Strategy. Action 7 under Target 2 of the EU Biodiversity Strategy through 2020 seeks to "ensure no net loss of biodiversity and ecosystem services". It is composed of two complementary sub-actions,[1] with sub-action 7b specifying that "the Commission will carry out further work with a view to proposing an initiative to ensure there is no net loss of ecosystems and their services by 2015 (e. g. through compensation or offsetting schemes)". In this respect, the previous detailed descriptions of legal and practical approaches to biodiversity offsetting in selected EU member states as well as the brief (non-exhaustive) discussion of other national approaches given here, will help fill this European initiative with new ideas and impulses.

[1] http://ec.europa.eu/environment/nature/biodiversity/nnl/index_en.htm

W. Wende (✉)
Leibniz Institute of Ecological Urban and Regional Development and TU Dresden, Dresden, Germany

National University of Singapore, Singapore, Singapore
e-mail: w.wende@ioer.de

W. Wende et al. (eds.), *Biodiversity Offsets*,
https://doi.org/10.1007/978-3-319-72581-9_13

Ecosystem Market Place points out that compensatory mitigation "is also used at a smaller scale in Denmark (at the municipal level), Finland (where a pilot is underway) [see also Darbi et al. 2010], and in Switzerland [see e.g. Dalang and Hersperger 2010] [. . .]."[2] Although Switzerland is a non-EU country, it repays consideration as a European nation with a long tradition in biodiversity offsetting. Finnish researchers have contributed a method to calculate minimum biodiversity offset multipliers (Laitila et al. 2014). Furthermore, an intensive discussion is currently being conducted in Finland on whether and how to implement habitat banking (Kniivilä et al. n.d.).

The Swedish system has also been well established for several years, with the country's scientists and practitioners conducting a living discussion on how best to implement biodiversity offsetting (Rundcrantz and Skaerbaeck 2003). One sector in the country that sees many offsetting measures is road building (Rundcrantz 2006, 2007; Koh et al. 2017). Ecosystem Market Place states: "Mitigation is typically carried out by the municipal government, and mitigation requirements may be relatively relaxed in terms of requiring like-for-like or spatially relevant compensation." Further information on biodiversity offsetting for all Nordic countries can be found in a study conducted by Enetjärn et al. (2015).

A new nature conservation law promulgated in Hungary in 1996 (Act. No. LIII of 1996 on Nature Conservation in Hungary) imposes an obligation to provide compensation for interventions in nature (Peters et al. 2003). Unfortunately, a wider investigation regarding the extent to which this law is applied or whether this obligation still exists in the year 2017 is not possible within the constraints of this book. Yet there are indications that compensatory measures are being applied in Hungary in the road building sector, resulting in the construction of wildlife bridges and fences to protect wildlife as well as the reintroduction or resettlement of protected species and animal communities (Peters et al. 2003, 158 f.).

Once again, we emphasize that the research conducted for this book could not encompass all EU member states. This would have required separate studies of every country in order to evaluate legal and planning documents drawn up in the local language. With this in mind, it is surprising to see how many detailed approaches can be found in most EU member states, also for specific areas of application. Furthermore, we cannot discount the existence of alternative approaches. Once again, we ask for the reader's forgiveness that some European approaches have been left out of discussion. This also underlines the need for further detailed study of existing offset measures in the individual member states, which the EU Commission could realise through research projects. The same is true of research into approaches adopted by countries outside the EU such as Australia, the USA or South Africa, to name but three (cf. Koh et al. 2014).

Finally, of course, all EU member states have to follow the requirements of the EU EIA Directive, which in paragraph 6 of Annex IV (Information Referred to in

---

[2] http://www.ecosystemmarketplace.com/marketwatch/biodiversity/europe/#denmark-finland-and-switzerland [last accessed 3 Mar 2017].

Article 5 (1) No. 6 for the Environmental Impact Study) demands "A description of the measures envisaged to prevent, reduce and where possible offset any significant adverse effects on the environment."[3] However, the extent to which these offset and compensation obligations imposed by the EIA are actually considered and implemented in practice is largely unknown. The latest amendment to the EU EIA Directive strengthens the obligation to consider and implement avoidance, mitigation and compensation liabilities. Thus in Article 8a of the new EU EIA Directive[4] we read that: "1. The decision to grant development consent shall incorporate at least the following information: [...] (b) [...] a description of any features of the project and/or measures envisaged to avoid, prevent or reduce and, if possible, offset significant adverse effects on the environment as well as, where appropriate, monitoring measures." Additionally, Article 8a No. 4 of the new EU EIA Directive states that: "In accordance with the requirements referred to in paragraph 1 (b), Member States shall ensure that the features of the project and/or measures envisaged to avoid, prevent or reduce and, if possible, offset significant adverse effects on the environment are implemented by the developer, and shall determine the procedures regarding the monitoring of significant adverse effects on the environment." The stricter requirement to undertake an EIA for public and private projects may already be encouraging the practice of offsetting. However, the fact that this is insufficient to achieve full no net loss of biodiversity and ecosystem services is shown by the approaches of member states that have established EIAs and yet have been forced to adopt more rigorous offsetting systems to implement biodiversity and ecosystem services measures on the ground in order to realise no net loss.

References

Dalang T, Hersperger AM (2010) How much compensation do we need? Replacement ratio estimates for Swiss dry grassland biotopes. Biol Conserv 143:1876–1884

Darbi M, Ohlenburg H, Herberg A, Wende W (2010) Impact mitigation and biodiversity offsets—compensation approaches from around the world. A study on the application of Article 14 of the CBD. Bundesamt für Naturschutz (Naturschutz und Biologische Vielfalt) Bonn

Enetjärn A, Cole S, Kniivilä M, Hårklau S-E, Hasselström L, Sigurdson T, Lindberg J (2015) Environmental compensation – key conditions for increased and cost effective application. Nordic Council of Ministers, Denmark. http://www.greengrowthknowledge.org/resource/environmental-compensation-key-conditions-increased-and-cost-effective-application. Last Accessed 10 Aug 2017

Kniivilä M, Kosenius A-K, Horne P (n.d.) Creating market for biodiversity by using habitat banking: preliminary assessment of applicability to Finland. http://ageconsearch.umn.edu/bitstream/199234/2/Kniivil%C3%A4,%20M.pdf. Last Accessed 3 Mar 2017

[3]Directive 2011/92/EU of the European parliament and of the council of 13 December 2011 on the assessment of the effects of certain public and private projects on the environment.

[4]Directive 2014/52/EU of the European parliament and of the council of 16 April 2014 amending Directive 2011/92/EU on the assessment of the effects of certain public and private projects on the environment.

Koh N-S, Hahn T, Ituarte-Lima C (2014) A comparative analysis of ecological compensation programs: the effect of program design on the social and ecological outcomes. Working paper Uppsala University. http://uu.diva-portal.org/smash/get/diva2:772933/FULLTEXT01.pdf. Last Accessed 6 Mar 2017

Koh N-S, Hahn T, Ituarte-Lima C (2017) Safeguards for enhancing ecological compensation in Sweden. Land Use Policy 64:186–199

Laitila J, Moilanen A, Pouzols F-M (2014) A method for calculating minimum biodiversity offset multipliers accounting for time discounting, additionality and permanence. Meth Ecol Evol 5 (11):1247–1254. https://doi.org/10.1111/2041-210X.12287

Peters W, Siewert W, Szaramowicz M (2003) Folgenbewältigung von Eingriffen im internationalen Vergleich. In: Bundesamt für Naturschutz (ed) BfN-Skripten No. 82. BfN-Eigenverlag, Bonn

Rundcrantz K (2006) Environmental compensation in Swedish road planning. Eur Environ 16 (6):350–367

Rundcrantz K (2007) Environmental compensation for disrupted ecological functions in Swedish road planning and design. Faculty of landscape planning, Horticulture and Agricultural Science Department of Landscape Architecture, Doctoral thesis Swedish University of Agricultural Science, Alnarp

Rundcrantz K, Skaerbaek E (2003) Environmental compensation in planning: a review of five different countries with major emphasis on the German system. Eur Environ 13:204–226

Conclusions: Lessons from Biodiversity Offsetting Experiences in Europe

Graham Tucker, Marianne Darbi, Wolfgang Wende, Fabien Quétier, and Matt Rayment

It is clear that Europe is still far from its target to halt the loss of biodiversity and ecosystem services (EEA 2010a, b; European Commission 2015), and this is in large part due to biodiversity impacts in the wider environment, i.e. outside protected areas. Many such impacts are the result of multiple small, scattered developments that individually often do not justify refusal on environmental grounds, but nonetheless have substantial incremental and cumulative impacts; thereby leading to 'death by 1000 cuts'. To tackle this problem and halt biodiversity losses, it is becoming increasingly evident that complementary measures to strict protection are required that have the realistic ambition of achieving no net loss, through the avoidance, minimisation and offsetting of residual impacts. Such no net loss/offset policies can form a key component of sustainable development in that they can help

G. Tucker (✉)
Institute for European Environmental Policy, London, UK
e-mail: GTucker@ieep.eu

M. Darbi
Helmholtz Centre for Environmental Research, Leipzig, Germany
e-mail: marianne.darbi@ufz.de

W. Wende
Leibniz Institute of Ecological Urban and Regional Development and TU Dresden, Dresden, Germany

National University of Singapore, Singapore, Singapore
e-mail: w.wende@ioer.de

F. Quétier
BIOTOPE, Mèze, France
e-mail: fquetier@biotope.fr

M. Rayment
Consultant, London, UK
e-mail: matt.rayment@outlook.com

W. Wende et al. (eds.), *Biodiversity Offsets*,
https://doi.org/10.1007/978-3-319-72581-9_14

to regulate and manage necessary trade-offs between economic development and the conservation of biodiversity and ecosystem services (Quétier et al. 2014).

Consequently, there are an increasing number of no net loss and offsetting policies and initiatives in several European countries. In addition to the legal requirements for offsetting of unavoidable residual impacts following a strict mitigation hierarchy under Art. 6.3 and 6.4 of the EU Habitats Directive, the EU biodiversity strategy includes a broader action aimed at ensuring no net loss of biodiversity and ecosystem services. In the development of no net loss policies and related legislative instruments, governance, implementation procedures and standards, many have drawn from the wide ranging international experience of offsetting outside the EU, such as in the USA, Australia and South Africa. Offsetting principles, standards and good practices have been identified, disseminated and often incorporated and promoted in initiatives such as BBOP. Whilst such experiences and the BBOP principles (see chapter "Principles of the Business and Biodiversity Offsets Programme" in this book) are undoubtedly of considerable relevance to offsetting in Europe, their implementation in the particular context of European countries would benefit from further interpretation and experience review.

The European context for offsetting is very different to many non-European countries where offsetting has been long established. In particular, very little of Europe now consists of natural pristine habitats, and instead it is dominated by semi-natural habitats in some areas or more widespread highly modified cultural landscapes with multiple uses and users. Whilst semi-natural habitats are of very high biodiversity value and often decline, more modified habitats are still of significant biodiversity value, especially in terms supporting ecosystem services. Furthermore, Europe has a relatively comprehensive biodiversity policy framework. Natura 2000, the largest network of protected areas in the world (more than 18% of land area) protects a substantial proportion of the remaining semi-natural habitats. Within the Natura 2000 network there are mandatory requirements for biodiversity offsetting following a strict application of the mitigation hierarchy. However, outside this and national protected area networks, biodiversity protection tends to be relatively weak and, as discussed above, incremental losses are widespread.

The above factors are of considerable relevance. Strictly controlled offsetting is already in place as a last resort, following rigorous application of the mitigation hierarchy for the most threatened habitats and species and vulnerable sites (i.e. the Natura 2000 network). Beyond this, most developments in the wider environment, affecting cultural landscapes and less threatened biodiversity are still only weakly regulated. Business as usual in these parts of Europe is leading to a continued loss of biodiversity and ecosystem services. Consequently, there are relatively low risks of offsetting further weakening biodiversity protection in the wider environment, while there is potentially much to be gained. This is supported by the findings in this book and other studies which have found little evidence of current offsetting schemes in Europe being abused as a 'licence to trash' (Tucker et al. 2014); although this is a potential risk to guard against. Furthermore, as much biodiversity outside the Natura 2000 network is associated with highly modified habitats, offsetting is often more practical and has lower risks of failure than for impacts on pristine habitats or threatened species, for which offsettability of impacts is justifiably disputed (Pilgrim et al. 2013). Of course, challenges to implementation remain. For example, the time required to deliver the offset outcome can be underestimated and not properly taken into account in metrics, while finding and securing suitable land can be difficult.

In addition, it is important to recognise that much of biodiversity in Europe is not viable in the long-term, as it is associated with increasingly fragmented and isolated patches of habitat that are subject to ongoing pressures (e.g. disturbance, pollution, high levels of predation) and continue declining. Appropriate conservation actions in the context of offsetting often include the consolidation of larger and more connected areas of habitat, rather than struggling to maintain isolated fragments. In such a context, the mitigation hierarchy needs to be applied particularly carefully, with due consideration of what avoidance and mitigation measures are appropriate, and weighing the potential biodiversity outcomes of on-site offsetting against off-site offsetting.

The factors discussed above indicate that while the development of no net loss policies and offsetting in Europe should certainly take on board international standards, principles and good practices, it should interpret them carefully and also take into account European experiences and lessons from decades of conservation successes and failures. However, offsetting experiences in Europe have been relatively little publicised outside a few countries (Austria, Germany, Sweden and Switzerland), where it has been most established, and information on more recent developments in other countries is fragmented (France, Spain etc.). Consequently, few people are aware that offsetting is actually taking place in parts of Europe (outside Natura 2000 sites), let alone aware of its effectiveness or the lessons that can be drawn from the schemes.

This book has attempted to pull together current European experience and has shown that this has much to offer in terms of informing debates on offsetting in Europe, and elsewhere. To help inform further debates on EU and national initiatives, the key lessons that we have drawn from the experiences reviewed in this book are summarised below.

- To complement and avoid undermining existing nature conservation policies and legislation, offsetting policies need to focus on habitats and species in the wider environment that are not already protected by the EU Birds and Habitats Directives or equivalent national measures. Where such offsetting policies are implemented it is essential that they do not directly or indirectly reduce the existing levels of protection of species and habitats and protected areas. This would be, quite justifiably, illegal within Natura 2000 sites under the Habitats Directive, and it is essential to maintain this high standard. This danger can be reduced by keeping offsetting schemes targeting the wider environment separate from measures directed towards Natura 2000 sites.
- Evidence clearly shows that for offsetting to take place at sufficient scale to have significant beneficial biodiversity impacts in the EU it needs to be based on unambiguous legislative requirements for the achievement of no net loss (or a net gain). This is demonstrated by the commonplace offsetting of biodiversity

impacts in the wider environment in Germany, Austria and France (see chapters "Germany", "Austria" and "France") that result from clear legal obligations. This contrasts with the minimal level of offsetting that occurs in the UK where legal requirements are unclear and they are variably interpreted, despite a recent initiative in England to promote offsetting (chapter "United Kingdom"). However, even in Germany, legislation is not enough by itself to guarantee implementation, as studies have shown low levels of implementation in the past, where monitoring and enforcement by authorities was weak.

- For offsetting schemes to be effective in terms of delivering long-term biodiversity benefits, experiences shows that they need to be based on well-established principles, standards and good practice (as described in chapter "Principles of the Business and Biodiversity Offsets Programme") interpreted in accordance with their European and local context. These need to be clearly and unambiguously incorporated into the legislative instruments requiring offsetting and associated rules and guidance. This helps to provide clarity for businesses/developers and offset providers so that they know what is required of them and that they are being treated in a fair and equal manner to other developers and offset providers. Clear regulation also enables authorities to monitor and enforce requirements, and for other stakeholders and the public to understand offsetting processes and obligations so that they can also raise the alarm if offsets do not meet their agreed requirements.
- As discussed above, offsetting must be carried out in accordance with the mitigation hierarchy, and care needs to be taken to ensure that all proposed avoidance, mitigation and offsetting measures are appropriate, proportionate and realistic with similar levels of scrutiny given to each stage of the mitigation hierarchy. Similarly, decisions on whether offsetting should be on-site or off-site need to be based on careful case by case consideration of what options provide the best and most reliable outcome for biodiversity and ecosystem services, rather than simplistic general rules. Social, health, security and economic issues should not be ignored but incorporated into the broader decision-making process, without jeopardising biodiversity outcomes.
- Another key challenge of ensuring no net loss and effective offsetting lies in providing real added value (i.e. additionality) compared to what would have happened in its absence. To achieve this, most obviously, offsets should not be placed within Natura 2000 sites or other protected areas where countries already have obligations to conserve and, if necessary, restore the key habitats and species of conservation concern within them. Unless it is obvious and beyond doubt that the offsets are providing benefits that are above and beyond those that are already required or planned for, this would amount to cost shifting and lead to a continued net loss of biodiversity from development. Required and planned commitments to protected areas therefore need to be monitored carefully to safeguard against a gradual lowering of their ambition, which might occur to accommodate the needs of developers for offsetting opportunities.
- A fundamental requirement for ensuring that offsets contribute to achieving no net loss of biodiversity is the use of appropriate metrics and exchange rules. This book shows that a wide variety of metrics are used in Europe, which can lead to

problems (e.g. comparability problems and associated reduced transparency and efficiency of use), and some are probably too simplistic and unreliable to ensure no net loss is achieved. Whilst there is no best type of metric, whichever is used should reflect the importance of the various biodiversity components that are impacted by the development, for example using appropriate habitat typologies and incorporating assessments of habitat condition as well as area. At the same time, however, it is important to ensure that metrics are workable and proportionate. Multipliers should also be used where appropriate to adjust metrics according to potential risks of offset underperformance and the need to compensate for the time required to biodiversity gain from the offset. Setting these multipliers is difficult, and therefore ex-post monitoring is required to validate or correct them.

- It is widely accepted good practice to combine metrics with appropriate exchange rules that take a precautionary approach, such as requiring like-for-like offsets for habitats of significant biodiversity value, because metrics do not capture all important biodiversity values. However, it will often be appropriate for exchange rules to encourage trading up of offsets, i.e. restoring higher biodiversity value habitats than those impacted when they are of low value. Exchange rules are also normally applied to the location of offsets relative to impacts, but as noted above close proximity is not always better.
- There is clear European and wider evidence that the potential benefits of biodiversity offsetting can be greatly increased through strategic landscape planning, whereby areas that would benefit most from habitat creation, recreation or restoration are identified. Such benefits can be further increased by the aggregation of offsets to create larger or better connected patches of habitat. The location and aggregation of offsets in such areas can be encouraged by purchase (or designation) of these areas by municipalities or other government authorities or agencies to offset public biodiversity impacts, and/or by habitat banks. Offsetting in such areas, and their aggregation, can also be encouraged by metrics that incorporate landscape ecology criteria.
- Clear legal requirements for the offsetting of losses of relatively widespread habitats and species can lead to sufficient offsetting activities to stimulate a "market" for biodiversity gains ("credits") from offsets, where several offset providers compete for offset implementation. In some countries, this has evolved into various forms of habitat "banking", where providers must produce the biodiversity gains before being able to sell them to developers (see chapters "France", "Germany", "Other EU Member States" and "Spain" for examples from Germany and France, and discussions in some of the Scandinavian countries and Spain). Such banks can be a useful tool for the implementation and long-term maintenance of offsets and can increase biodiversity gains (e.g. by creating larger areas of habitat) and reduce offset risks by demonstrably creating habitats in advance of impacts. However, they may be less suitable e.g. for offsetting impacts on certain ecosystem services which are context and location specific. Thus, they bring policy and practical risks of their own, which need to be addressed through appropriate legislation and guidance, governance and enforcement.

- As biodiversity offsetting is a complex process with significant risks of failure, experience shows that adequate monitoring of each offset is needed by the competent authority to ensure its objectives are achieved, or additional enforcement and/or contingency measures taken if this is not the case. To obtain credibility and stakeholder support, full transparent reporting of the immediate and long-term achievements of offsets is necessary. However, it is apparent that such monitoring is variable across Europe. Furthermore, overall evaluations of the effectiveness and cost efficiency of offsetting policies are few and not very reliable, partly due to the ever-changing nature of policies but also to a lack of political willingness or capacity to monitor and publish results.
- Effective offsetting schemes take considerable time to be designed, planned and fully implemented, and significant ongoing resources are required for the authorities (or allocated third parties) responsible for their permitting, monitoring, reporting and if necessary enforcement, as well as, ideally, strategic planning in terms of identification of optimal offset location. However, in Europe there is wide evidence of underfunding of environmental authorities, which reduces their effectiveness in environmental protection, and may often be counter-productive in wider economic terms as slow or poor decision making can result in delays for developers and inefficient offsetting. This was evident in the UK, where capacity constraints in environmental and local authorities hindered the recent pilot offsetting initiative (Chapter "United Kingdom"). It is therefore clear that countries aiming to develop and implement offset policies recognize that this requires adequate increases in resources for permitting and enforcement authorities.
- Evidence shows that well planned and executed, and sufficiently ambitious offsets (which for example incorporate appropriate multipliers for offset risk and time delays) can result in substantial biodiversity gains, especially when they are strategically located. However, it is inappropriate to consider these as being a form of biodiversity funding as this can lead to conflicts of interests, within environmental authorities and NGOs. Furthermore, whilst some schemes may over deliver and provide substantial gains, others may under deliver or fail completely. As comprehensive and careful monitoring and accounting is not undertaken in any European country, it is uncertain whether offset policies provide overall no net loss, let alone net gains.
- An obvious gap in the no net loss policies and requirements for offsets in all EU countries (as indeed in many other parts of the world) is in the general lack of requirements for residual impacts from agricultural and forestry land uses, developments and improvements, and fisheries, to be subject to offsetting. As intensive practices in these sectors exert some of the highest and most widespread pressures on biodiversity, which are not within the scope of current no net loss initiatives, this results in a major policy gap which fundamentally undermines sustainable development and the achievement of the EU's and national biodiversity targets. Whilst it may be difficult and inefficient (e.g. with high administrative burdens) to deal with such impacts at a local level (e.g. for each farm) policy measures might be able to achieve no net loss at a larger-scale for each sector (Tucker et al. 2014).

Given these lessons, and the broader experiences of offsetting in Europe reviewed in this book, an obvious last question to consider is what could be the value of an EU level initiative on no net loss. One could argue that countries can set up effective no net loss policies and offsetting systems, without the need for an EU level framework, such as seen in some European countries (Darbi et al. 2010). Indeed it has been suggested that the German approach could form a good model for application elsewhere in Europe (Albrecht et al. 2014).

However, in most of Europe, offsetting outside the Natura 2000 network appears to be at low levels, and primarily a voluntary activity. In the absence of an EU level framework setting no net loss objectives for impacts in the wider environment, it is understandable that Member States will be reluctant to do more than some others, as offsetting is seen as increasing regulatory burdens on developers (often mistakenly) and it can have significant direct and indirect costs. Such voluntary approaches have been shown to be ineffective as offsets are only likely to be carried out by developers where their costs are insignificant.

Therefore, to help advance offsetting as a key element of sustainable development, it appears necessary to create a clearer EU level no net loss policy requirement, with associated standards and guidance, which could also help reduce burdens and costs. The creation of a level playing field may also enable Member States and business to advance offsetting ambitions and schemes without suffering a competitive disadvantage. An EU framework requiring the extension of offsetting to the agriculture, forestry and fishery sectors could greatly increase member states' abilities to address biodiversity declines in the wider environment.

Furthermore, an EU level framework may also ensure that the wider development of offsetting systems across Europe robustly incorporate the key principles, standards and best practices of offsetting. This is necessary because evidence shows that there are risks from introducing offsetting systems because poorly designed and weakly implemented and enforced offsetting can lead to significant biodiversity losses. Whilst evidence of such counter-productive offsetting in Europe appears to be lacking, this book has revealed that best practices are not being followed in some cases (e.g. the lack of exchange rules in Austria, the initial and therefore inexperienced habitat banking practices in Spain, the simplistic metric in the UK, the confusing variety of approaches in Germany and its federal states, and finally the limited long-term security of offsets in most countries). Yet all these initiatives claim to use offsets and to aim for no net loss, creating a confusing landscape for developers, conservation organizations and other stakeholders, including the public.

A common EU framework for offsetting could usefully set out key principles and minimum standards to be followed, and include tools to share information and promote best practices. This would help to ensure offsets are effective, equitable and long-lasting and thereby reliably achieve no net loss of biodiversity as a minimum and where possible, net gains. Without such a framework that sets high standards and a level playing field, a simple requirement for EU Member States to introduce no net loss policies and offsetting could actually undermine existing standards and encourage a race to the bottom. Therefore, whilst the development of a robust EU framework for the achievement of no net loss of biodiversity may be a challenge, there is good reason to believe that it would provide the political and policy basis for an increase in both the amount of offsetting and its effectiveness. Such an initiative therefore has the potential to make a significant contribution to reducing the incremental losses of biodiversity from developments in the wider environment that are currently being ignored.

References

Albrecht J, Schumacher J, Wende W (2014) The German impact mitigation regulation: a role model for a no net loss strategy and biodiversity offsets for halting the loss of biodiversity in the European Union? Environ Policy Law 44(3):317–325

Darbi M, Ohlenburg H, Herberg A, Wende W (2010) Impact mitigation and biodiversity offsets—compensation approaches from around the world. Naturschutz und Biologische Vielfalt 101, Bundesamt für Naturschutz, Bonn – Bad Godesberg

EEA (2010a) Assessing biodiversity in Europe – the 2010 Report. EEA Technical Report No 5/2010, European Environment Agency, Copenhagen

EEA (2010b) EU 2010 biodiversity baseline. EEA Technical Report No 12/2010, European Environment Agency, Copenhagen

European Commission (2015) The mid-term review of the EU biodiversity strategy to 2020, COM (2015) 478 final

Pilgrim JD, Brownlie S, Ekstrom JMM, Gardner TA, von Hase A, ten Kate K, Savy CE, Stephens RTT, Temple HJ, Treweek JR, Ussher GT, Ward G (2013) A process for assessing offsetability of biodiversity impacts. Conserv Lett 6(5):376–384

Quétier F, Regnery B, Levrel H (2014) No net loss of biodiversity or paper offsets? – a critical review of the French no net loss policy. Environ Sci Pol 38:120–131

Tucker GM, Allen B, Conway M, Dicki I, Hart K, Rayment M, Schulp CJE (2014) Policy options for an EU no net loss Initiative. Report to the European Commission (with Annexes), Institute for European Environmental Policy, London/Brussels

MIX
Papier aus verantwortungsvollen Quellen
Paper from responsible sources
FSC® C105338

If you have any concerns about our products,
you can contact us on
ProductSafety@springernature.com

In case Publisher is established outside the EU,
the EU authorized representative is:
Springer Nature Customer Service Center GmbH
Europaplatz 3, 69115 Heidelberg, Germany

Printed by Libri Plureos GmbH
in Hamburg, Germany